Thomas W. Wieting

Annals of Mathematics Studies

Number 53

ANNALS OF MATHEMATICS STUDIES

Edited by Robert C. Gunning, John C. Moore, and Marston Morse

1. Algebraic Theory of Numbers, *by* HERMANN WEYL
3. Consistency of the Continuum Hypothesis, *by* KURT GÖDEL
11. Introduction to Nonlinear Mechanics, *by* N. KRYLOFF *and* N. BOGOLIUBOFF
16. Transcendental Numbers, *by* CARL LUDWIG SIEGEL
19. Fourier Transforms, *by* S. BOCHNER *and* K. CHANDRASEKHARAN
20. Contributions to the Theory of Nonlinear Oscillations, Vol. I, *edited by* S. LEFSCHETZ
21. Functional Operators, Vol. I, *by* JOHN VON NEUMANN
24. Contributions to the Theory of Games, Vol. I, *edited by* H. W. KUHN *and* A. W. TUCKER
25. Contributions to Fourier Analysis, *edited by* A. ZYGMUND, W. TRANSUE, M. MORSE, A. P. CALDERON, *and* S. BOCHNER
28. Contributions to the Theory of Games, Vol. II, *edited by* H. W. KUHN *and* A. W. TUCKER
29. Contributions to the Theory of Nonlinear Oscillations, Vol. II, *edited by* S. LEFSCHETZ
30. Contributions to the Theory of Riemann Surfaces, *edited by* L. AHLFORS *et al.*
33. Contributions to the Theory of Partial Differential Equations, *edited by* L. BERS, S. BOCHNER, *and* F. JOHN
34. Automata Studies, *edited by* C. E. SHANNON *and* J. MCCARTHY
36. Contributions to the Theory of Nonlinear Oscillations, Vol. III, *edited by* S. LEFSCHETZ
38. Linear Inequalities and Related Systems, *edited by* H. W. KUHN *and* A. W. TUCKER
39. Contributions to the Theory of Games, Vol. III, *edited by* M. DRESHER, A. W. TUCKER *and* P. WOLFE
40. Contributions to the Theory of Games, Vol. IV, *edited by* R. DUNCAN LUCE *and* A. W. TUCKER
41. Contributions to the Theory of Nonlinear Oscillations, Vol. IV, *edited by* S. LEFSCHETZ
42. Lectures on Fourier Integrals, *by* S. BOCHNER. In preparation
43. Ramification Theoretic Methods in Algebraic Geometry, *by* S. ABHYANKAR
44. Stationary Processes and Prediction Theory, *by* H. FURSTENBERG
45. Contributions to the Theory of Nonlinear Oscillations, Vol. V, CESARI, LASALLE, *and* LEFSCHETZ
46. Seminar on Transformation Groups, *by* A. BOREL *et al.*
47. Theory of Formal Systems, *by* R. SMULLYAN
48. Lectures on Modular Forms, *by* R. C. GUNNING
49. Composition Methods in Homotopy Groups of Spheres, *by* H. TODA
50. Cohomology Operations, *lectures by* N. E. STEENROD, *written and revised by* D. B. A. EPSTEIN
51. Lectures on Morse Theory, *by* J. W. MILNOR
52. Advances in Game Theory, *edited by* M. DRESHER, L. SHAPLEY, *and* A. W. TUCKER
53. Flows on Homogeneous Spaces, *by* L. AUSLANDER, L. GREEN, F. HAHN, *et al.*
54. Elementary Differential Topology, *by* J. R. MUNKRES
55. Degrees of Recursive Unsolvability, *by* G. E. SACKS. In preparation

FLOWS ON HOMOGENEOUS SPACES

BY

L. AUSLANDER, L. GREEN, and F. HAHN

with the assistance of

L. MARKUS and W. MASSEY

and an Appendix by

L. GREENBERG

PRINCETON, NEW JERSEY

PRINCETON UNIVERSITY PRESS

1963

L. C. Card 62-19959

Printed in the United States of America

INTRODUCTION

In classical dynamics the action of a one-parameter group — time — on the phase space is one of the principal objects of study. Ergodic theory concerns itself with the case when the phase space is a measure space. Topological dynamics deals with topological phase spaces. Requiring that the phase space have both structures, and that they be compatible, brings us closer to the original dynamical problem. The questions asked about such flows include the following:

Topological transitivity: is there a dense orbit?

Minimality: is every orbit dense?

Ergodicity: are there any proper invariant subsets of positive measure?

Even asking that the phase space be a manifold and the measure locally Lebesgue still leads to very general, and generally unsolved, problems. In this book we consider the case when the manifold is a homogeneous space of a Lie group. Furthermore, the one-parameter group is assumed to be a subgroup of the original group, acting in a natural fashion. This enables one to get a firmer grip on the behavior of the flow, since it is now part of the action of a much larger group on the phase space. As in classical and quantum dynamics, the presence of these extra "symmetries" helps the analysis a great deal. In particular, the theory of group representations can be used with success.

During the academic year 1960-1961, The National Science Foundation sponsored a conference on Analysis in the Large at Yale University. This conference brought together for an extended period of time mathematicians from various parts of the country with different backgrounds. All but the last of the papers in this book have come directly from the interaction of the people at this conference. Markus and Hahn started things off with a search for examples of minimal flows on manifolds. Auslander suggested that flows on nilmanifolds induced by one parameter groups might be minimal. The results of these investigations are contained in Chapter IV and served as a starting point for almost all of the further material in this book. As an example of the direct advantage to the authors of the nature of the conference, it was Gottschalk who suggested the study of the distal nature of flows on nilmanifolds, which plays such a central role in the material in Chapter IV.

The methods used in Chapter IV, although capable of giving new information seemed to be incapable of generalization. Green arrived at this time for the second semester and, hearing of the results in Chapter IV, suggested that a study of the ergodic nature of flows on nilmanifolds might be possible by means of group representation theory. This program and its generalization to certain solvmanifolds is carried out in Chapters V and VII. In Chapter VI, it is shown how results on flows on nilmanifolds can be used to prove facts about flows on certain types of solvmanifolds. Chapter III contains a detailed account of the general state of affairs for one-parameter groups acting on compact three-dimensional homogeneous spaces obtained by identifying any connected, simply connected non-compact Lie group under a discrete subgroup. In order to make the material as self contained as possible, Chapters I and II were added as a listing and discussion of some of the results already in the literature with which the reader may not be familiar.

Chapter VIII is devoted to an application of the previous results concerning flows on nilmanifolds to diophantine approximations. In an Appendix, Greenberg shows how the classical work of Hedlund on minimal flows can be extended and he relates his results to the work of Mahler [1] on diophantine approximations.

The authors would also like to take this opportunity to thank Professors W. H. Gottschalk, G. A. Hedlund, S. Kakutani and J. C. Oxtoby for the many stimulating conversations which contributed so much to the material in this book. We would also like to acknowledge the financial support given by The National Science Foundation's Grants 15565, 11287 and the Office of Ordnance Research Contract DA-19-020-ORD-5254.

CONTENTS

FLOWS ON HOMOGENEOUS SPACES

CHAPTER I.

AN OUTLINE OF RESULTS ON SOLVMANIFOLDS

by

Louis Auslander

Introduction

The purpose of this chapter is to give an outline of the basic theorems in the theory of solvmanifolds. We will assume in presenting this material that the reader already has a basic knowledge of Lie groups and Lie algebras and the inter-relations between these two mathematical objects. Specifically, we will assume a familiarity with the material in either Chevalley [1] or Cohn [1]. With this as our starting point, we will outline the basic results in the theory of solvmanifolds. Since all this material is already in the literature, we will present no proofs, but will content ourselves with giving references.

We will adopt the following conventions in this paper. A Lie group need not be connected and will be denoted by $\underline{G}$, $\underline{S}$, $\underline{N}$, etc. We will call a connected Lie group an analytic group and generally use the same notation as for a Lie group. We will use $\underline{G}_0$ to denote the identity component of the Lie group $\underline{G}$. The Lie algebra of a Lie group will, of course, refer to the Lie algebra of the identity component of the Lie group and, if $\underline{G}$ denotes a Lie group, we will use $L(\underline{G})$ to denote the Lie algebra of $\underline{G}$. We will use $\exp(L(\underline{G}))$ to denote the exponential mapping of $L(\underline{G})$ to $\underline{G}$. Occasionally we will use the symbol L to denote a Lie algebra without specifying any Lie group. For W_1 and W_2 subalgebras of L, $[W_1,W_2]$ will denote the subalgebra generated by the bracket of an element in W_1 with an element of W_2. Analogously, if G is any group and H_1 and H_2 are subgroups, we will use $[H_1,H_2]$ to denote the group generated by all elements $h_1h_2h_1^{-1}h_2^{-1}$ for $h_1 \in H_1$ and $h_2 \in H_2$.

<u>Definition</u>: Let L be a Lie algebra and let

$$L_1 = L, \quad L_2 = [L_1,L_1],\ldots,L_k = [L_{k-1},L_{k-1}].$$

L is called solvable if $L_k = 0$ for some k.

Let L be a Lie algebra and let

$$L^1 = L, \quad L^2 = [L,L^1],\ldots,L^k = [L,L^{k-1}].$$

L is called nilpotent if $L^k = 0$ for some k.

The basic examples of these types of Lie algebras are the following: Lie algebras of matrices of the form

$$\begin{pmatrix} 0 & & * \\ & \ddots & \\ 0 & & 0 \end{pmatrix}$$

are all nilpotent.

Lie algebras of matrices of the form

$$\begin{pmatrix} a_1 & & * \\ & \ddots & \\ 0 & & a_n \end{pmatrix}$$

are all solvable.

<u>Definition</u>: Let G be a group and let

$$G_1 = G, \quad G_2 = [G_1, G_1], \ldots, G_k = [G_{k-1}, G_{k-1}].$$

Then G is called solvable if $G_k = e$ for some k.

Let G be a group and let

$$G^1 = G, \quad G^2 = [G^1, G^1], \ldots, G^k = [G^1, G^{k-1}].$$

If $G^k = e$ for some k, then G is called nilpotent. Clearly if G is nilpotent it is solvable.

<u>Remark</u>: The homomorphic image of a solvable (nilpotent) group is solvable (nilpotent). A subgroup of a solvable (nilpotent) group is solvable (nilpotent).

<u>Example</u>: Any group of matrices of the form

$$\begin{pmatrix} 1 & & * \\ & \ddots & \\ 0 & & 1 \end{pmatrix}$$

is nilpotent.

Any group of matrices of the form

$$\begin{pmatrix} a_1 & & * \\ & \ddots & \\ 0 & & a_n \end{pmatrix}$$

is solvable.

1. Basic Facts.

I. Let $L(\underline{G})$ be the Lie algebra of the analytic group $\underline{G}$. Then

a) $L(\underline{G})$ is solvable if and only if $\underline{G}$ is solvable.

b) $L(\underline{G})$ is nilpotent if and only if $\underline{G}$ is nilpotent.

c) If L is a solvable Lie algebra, then $[L,L]$ is a nilpotent ideal in L. Further, there exists a unique maximal nilpotent ideal M in L.

II. Let $\underline{G}$ be an analytic, simply connected, solvable group. Then $\underline{G}$ is homeomorphic to euclidean space. (Chevalley [2].)

III. a) Let $\underline{G}$ be an analytic, simply connected, solvable group. Then $[\underline{G},\underline{G}]$ is an analytic, simply connected, normal subgroup of $\underline{G}$ and $[\underline{G},\underline{G}]$ is nilpotent. (Ibid.)

b) If $\underline{G}$ is an analytic, simply connected, solvable group, then $\underline{G}$ has a unique maximal analytic normal nilpotent subgroup $\underline{M}$.

c) If $L(\underline{G})$ is the Lie algebra of $\underline{G}$, then $M = L(\underline{M})$, where M is the maximal nilpotent ideal of $L(\underline{G})$ and $\underline{M}$ is the maximal analytic normal nilpotent subgroup of $\underline{G}$.

IV. Every representation of a solvable Lie algebra over the complex numbers is equivalent to a representation of the form

$$\begin{pmatrix} a_1 & & * \\ & \ddots & \\ 0 & & a_n \end{pmatrix}$$

(Jacobson [1].)

V. Every Lie algebra has a faithful matrix representation. Every nilpotent Lie algebra has a faithful matrix representation of the form

$$\begin{pmatrix} 0 & & * \\ & \ddots & \\ 0 & & 0 \end{pmatrix}$$

(Jacobson [1].)

Remark: Not every representation of a nilpotent Lie algebra is equivalent to one of the above form.

2. The Exponential Mapping in Analytic Solvable Groups.

Henceforth analytic solvable or nilpotent groups will always be assumed to be simply connected, unless explicitly stated to the contrary.

I. If $\underline{G}$ is a nilpotent analytic group, then the mapping

$$\exp\,(L(\underline{G})) \longrightarrow \underline{G}$$

is a homeomorphism of $L(\underline{G})$ onto $\underline{G}$. (Malcev [1].)

II. If $\underline{G}$ is a solvable analytic group the exponential mapping need not be onto or a homeomorphism as the following example shows. Let (r,s,t) be three real parameters and let $\underline{S}^3$ be the three-dimensional analytic group

$$\begin{pmatrix} \cos 2\pi t & \sin 2\pi t & 0 & r \\ -\sin 2\pi t & \cos 2\pi t & 0 & s \\ 0 & 0 & 1 & t \\ 0 & 0 & 0 & 1 \end{pmatrix}$$

$L(S^3)$ is given by the triple (a,b,c) representing the element

$$\begin{pmatrix} 0 & \pi a & 0 & b \\ -2\pi a & 0 & 0 & c \\ 0 & 0 & 0 & a \\ 0 & 0 & 0 & 0 \end{pmatrix}$$

It is straight forward to verify that $\exp(1,b,c) = (0,0,1)$ if (r,s,t) are taken as a coordinate system in $\underline{S}^3$. Similarly the points $(r,s,1)$ for $r^2 + s^2 \neq 0$, are not in the image of $L(\underline{S}^3)$ under the exponential mapping. $\underline{S}^3$ is characteristic of the type of pathology which can occur in the exponential mapping of solvable analytic groups in the following sense:

III. Let $\underline{S}$ be a solvable analytic group. The exponential mapping is a homeomorphism of $L(\underline{S})$ onto $\underline{S}$ if and only if there is no homomorphism of $\underline{S}$ onto $\underline{S}^3$. Dixmier [2].

IV. Let $\underline{G}$ be an n dimensional solvable analytic group. Then there exist n one-parameter subgroups $g_i(t)$, $i = 1,\ldots,n$ such that:

1. For $g \in \underline{G}$, $g = g_1(t_1)\ldots g_n(t_n)$ and this representation is unique where the right hand side denotes group multiplication.

2. If $\underline{G}_i = \{g_i(t_i)\ldots g_n(t_n) \mid t_i,\ldots,t_n$ any real no.$\}$, then $\underline{G}_i$ is a closed subgroup of $\underline{G}$ and $\underline{G}_i$ is normal in $\underline{G}_{i-1}$. (Iwasawa [1].)

3. Measures and Solvmanifolds.

Definition: Let $\underline{G}$ be a solvable analytic group and let $\underline{H}$ be a closed subgroup. Then $\underline{G}/\underline{H}$ is called a solvmanifold.

I. Haar measure on $\underline{G}$ induces a measure on $\underline{G}/\underline{H}$. (Mostow [2].)

II. If the total measure of $\underline{G}/\underline{H}$ is finite in any of the above measures, then $\underline{G}/\underline{H}$ is compact. (Ibid.)

III. If $\underline{G}/\underline{H}$ is compact, then G has a left and right invariant measure. (Ibid.)

Henceforth we will restrict ourselves to compact solvmanifolds.

Definition: Let $\underline{G}$ be a nilpotent analytic group and let $\underline{H}$ be a closed subgroup. Then $\underline{G}/\underline{H}$ is called a nilmanifold.

4. Malcev's Results on Nilmanifolds.

I. Let $\underline{N}$ be a nilpotent analytic group and let $\underline{H}$ be a closed subgroup such that $\underline{N}/\underline{H}$ is compact. Then if $\underline{H}_o$ denotes the identity component of $\underline{H}$, $\underline{H}_o$ is a normal subgroup of $\underline{N}$ and if $\underline{N}^* = \underline{N}/\underline{H}_o$ and $\Gamma = \underline{H}/\underline{H}_o$, we have:

a) $\underline{N}/\underline{H}$ is homeomorphic to $\underline{N}^*/\Gamma$.

b) $\underline{N}^*$ is a simply connected, analytic nilpotent group.

c) Γ is a discrete subgroup of $\underline{N}^*$.

d) Γ is the fundamental group of $\underline{N}/\underline{H}$.

Henceforth we will restrict the discussion of nilmanifolds to the case of $\underline{N}/\Gamma$, where $\underline{N}$ is an analytic nilpotent group and Γ is a subgroup such that $\underline{N}/\Gamma$ is compact.

II. $\Gamma \cap [\underline{N},\underline{N}] = \Gamma_1$ is a discrete subgroup of $[\underline{N},\underline{N}]$ and $[\underline{N},\underline{N}]/\Gamma_1$ is compact.

III. Existence of Malcev coordinates. Let $\underline{N}$ be an n dimensional nilpotent analytic group and let Γ be a discrete subgroup such that

$\underline{N}/\Gamma$ is compact. Then there exist n one-parameter groups $x_i(t)$ such that

$$\underline{N} = \{x_1(t_1)\dots x_n(t_n)\}$$

satisfying the conclusions of 3 IV, and such that

$$\Gamma = \{x_1(m_1)\dots x_n(m_n)\}$$

where the m_i are integers, $i = 1,\dots,n$.

IV. A necessary and sufficient condition for a group Γ to be the fundamental group of a compact nilmanifold is that

a) Γ contain no elements of finite order.

b) Γ is finitely generated.

c) Γ is nilpotent.

V. Let $\underline{N}_1$ and $\underline{N}_2$ be analytic nilpotent groups and let Γ_1 and Γ_2 be discrete subgroups of $\underline{N}_1$ and $\underline{N}_2$, respectively, such that $\underline{N}_1/\Gamma_1$ and $\underline{N}_2/\Gamma_2$ are compact. Let

$$\begin{array}{ccc} \Gamma_1 & \xrightarrow{\ i\ } & \underline{N}_1 \\ \downarrow{\scriptstyle\alpha} & & \downarrow{\scriptstyle\beta} \\ \Gamma_2 & \longrightarrow & \underline{N}_2 \end{array}$$

where i denotes the injection mappings in both groups and α is an isomorphism of Γ_1 onto Γ_2. Then there exists a unique isomorphism β of $\underline{N}_1$ onto $\underline{N}_2$ such that the diagram is commutative.

Corollary 1: If Γ is a discrete subgroup of the analytic group $\underline{N}$ such that $\underline{N}/\Gamma$ is compact, then every automorphism of Γ is uniquely extendable to an automorphism of $\underline{N}$.

Corollary 2: If M_1 and M_2 are compact nilmanifolds and $\pi_1(M_1)$ is isomorphic to $\pi_1(M_2)$, then M_1 is homeomorphic to M_2.

VI. A necessary and sufficient condition that an analytic nilpotent Lie group $\underline{N}$ have a discrete subgroup Γ such that $\underline{N}/\Gamma$ is compact is that $L(\underline{N})$ have rational constants of structure relative to some basis.

Remark: Since there exist nilpotent Lie groups such that the constants of structure of $L(\underline{N})$ are not rational relative to any basis, not every nilpotent Lie group $\underline{N}$ has a discrete subgroup Γ such that $\underline{N}/\Gamma$ is compact.

5. Solvmanifolds.

In solvmanifolds the immediate generalizations of 4.I and 4.V are false. Since these served as the basic tool theorems in the study of nilmanifolds, the study of solvmanifolds has been much more involved than the study of nilmanifolds. The first major results were the following of Mostow [1].

I. Let $\underline{S}$ be a solvable analytic group and $\underline{C}$ a closed subgroup of $\underline{S}$

such that $\underline{S}/\underline{C}$ is compact. Further, let $\underline{N}$ be the maximal analytic normal nilpotent subgroup of $\underline{S}$. Assume that $\underline{C}$ contains no proper normal analytic subgroup of $\underline{S}$. Then $\underline{CN}$ is closed and the connected component of the identity in $\underline{C}$ is a normal subgroup in $\underline{N}$.

II. Let M_1 and M_2 be two compact solvmanifolds and let $\pi_1(M_1)$ and $\pi_1(M_2)$ denote the fundamental groups of M_1 and M_2 respectively. Then if $\pi_1(M_1)$ is isomorphic to $\pi_1(M_2)$, M_1 is homeomorphic to M_2.

III. Let M_1 be a compact solvmanifold. Then M_1 is the bundle space of a fiber bundle over a torus with a nilmanifold as fiber.

Corollary: If M is a compact solvmanifold and $\pi_1(M)$ is the fundamental group of M, then $\pi_1(M)$ satisfies the exact sequence

$$1 \longrightarrow \Delta \longrightarrow \Pi_1(M) \longrightarrow Z^s \longrightarrow 1$$

where Δ is the fundamental group of a compact nilmanifold and Z^s is the integers taken s times.

The next major study of solvmanifolds was that of H. C. Wang [1]. His results may be summarized as follows:

IV. Let Γ satisfy the exact sequence

$$1 \longrightarrow \Delta \longrightarrow \Gamma \longrightarrow Z^s \longrightarrow 1$$

where Δ is the fundamental group of a compact nilmanifold and Z^s is the integers taken s times. Then there exists an analytic solvable Lie group $\underline{S}$ containing Γ as a discrete subgroup. Of course, $\underline{S}/\Gamma$ need not be compact.

V. Given Γ satisfying the hypothesis of IV above, there exists a Lie group $\underline{S}$ (not necessarily connected, but simply connected) such that $\underline{S} \supset \Gamma$ as a discrete subgroup with $\underline{S}/\Gamma$ compact. Further, if $\underline{S}_o$ denotes the identity component of $\underline{S}$, we have $\underline{S} = F \cdot \underline{S}_o$, where F is a finite abelian group and the dot denotes the semi-direct product.

The author (Auslander [1]), using the above, was able to prove

VI. Every fundamental group of a solvmanifold is the fundamental group of a compact solvmanifold.

Let $\underline{S}$ be an analytic solvable group and let Γ be a discrete subgroup of $\underline{S}$ such that $\underline{S}/\Gamma$ is compact. Then the author has characterized the groups Γ that satisfy this condition. Since we have no need for this result in the rest of the papers, and it is reasonably involved, we will not give it here.

6. Solvable Lie Groups Acting on Compact Nilmanifolds.

I. Let $\underline{S}$ be an analytic solvable group and let Γ be a discrete subgroup of $\underline{S}$ such that $\underline{S}/\Gamma$ is compact. Further, let Γ be a discrete subgroup of the analytic nilpotent group $\underline{N}$ such that $\underline{N}/\Gamma$ is compact. Then there exists a torus group $\underline{T}$ (compact abelian group) of automorphisms of $\underline{N}$ such that $\underline{S} \subset \underline{T} \cdot \underline{N}$, where the dot denotes the semidirect product. Further, if $\underline{H}$ is the maximal normal nilpotent analytic subgroup of $\underline{S}$, then $\underline{H} \subset \underline{N}$ and $\underline{T}$ acts trivially on $\underline{N}/\underline{H}$. (Auslander [2].)

CHAPTER II.

ERGODIC THEORY AND GROUP REPRESENTATIONS

by

L. Green

§1. The first explicit use of infinite-dimensional group representations in ergodic theory was by Gelfand and Fomin [1], who noticed that the geodesic flow on surfaces of constant negative curvature is an instance of the action of a one-parameter subgroup on a homogeneous (actually double-coset) space. Their method depended on knowing all the irreducible representations of the group, and examining in detail their restrictions to the one-parameter subgroup of the flow. Mautner [1] pointed out that for this sort of application it is not necessary to know every representation, but only the representations of sufficiently many subgroups. We shall use both of these methods.

Recently several Russian mathematicians have exploited Kolmogorov's concept of entropy, not only to prove ergodicity and mixing for certain flows but also to obtain detailed information about the spectra. Of the cases we investigate, only the geodesic flow (Chapter III, Theorem 5.3a) can be treated this way. The horocycle flow has zero entropy, and this is probably true of all the other flows we deal with. (For a description of results on entropy and for further references, see Rohlin [1].)

This chapter is intended to bring together known theorems and definitions in order to make the later exposition as self-contained as possible. The only results which have not appeared in print before are Theorem C and its corollary.

In §2 we review definitions from ergodic theory. §3 introduces enough group representation terminology to enable us to describe the application to ergodic theory. Some definitions and theorems dealing with induced representations are stated in §4. Finally, in §5, we obtain all irreducible unitary representations of two specific groups used in the sequel.

§2. Consider a measure space (S,Σ,μ), where Σ is a σ-field of subsets of S, and $\mu(S) < \infty$. An <u>automorphism</u> φ of (S, Σ, μ) is an invertible map of S onto itself which takes Σ onto itself such that $\mu(\varphi(E)) = \mu(E)$, for every $E \in \Sigma$. A <u>measure-preserving flow</u> on (S,Σ,μ) is a homomorphism $t \to \varphi_t$ of the reals into the group of automorphisms. The flow is called <u>continuous</u> if, for any elements A and B in Σ,

$\mu(\varphi_t(A) \cap B)$ is a continuous function of t. The flow is ergodic if $\varphi_t(E) = E$ for all t implies $\mu(E)\mu(S-E) = 0$. (Here it is assumed, of course, that $E \in \Sigma$. In what follows, this condition will be understood.) It turns out that φ_t is ergodic if and only if, for every A and B the Cesaro means

$$\frac{1}{T}\int_0^T \mu(\varphi_t(A) \cap B)\, dt$$

converge to $\mu(A)\mu(B)/\mu(S)$. If this convergence is also absolute, i.e., if

$$\lim_{T \to \infty} \frac{1}{T}\int_0^T \left|\mu(\varphi_t(A) \cap B) - \frac{\mu(A)\mu(B)}{\mu(S)}\right| dt = 0 ,$$

then φ_t is called weakly mixing. Finally, if

$$\lim_{T \to \infty} \mu(\varphi_T(A) \cap B) = \mu(A)\mu(B)/\mu(S) ,$$

φ_t is called strongly mixing.

A continuous flow φ_t defines a one-parameter group of unitary transformations U_t in $L_2(S,\Sigma,\mu)$ by the equation

$$(1) \qquad (U_t f)(s) = f(\varphi_t(s)), \qquad f \in L_2(S) .$$

The ergodic and mixing properties of φ_t are reflected in the spectrum of U_t. By the spectrum of U_t we mean the spectrum of the infinitesimal generator of U_t; or, more directly, the spectrum associated with the resolution of the identity $E(\Delta)$ for which, according to Stone's theorem,

$$(2) \qquad U_t = \int_{-\infty}^{\infty} e^{it\lambda}\, E(d\lambda) .$$

Since φ_t is ergodic if and only if the only invariant functions are constant, ergodicity is equivalent to $\lambda = 0$ having multiplicity one in the spectrum of U_t. φ_t is weakly mixing if and only if φ_t is ergodic and the point spectrum of U_t consists of zero alone (i.e., the remaining spectrum is continuous). φ_t is strongly mixing if and only if, for any $f,g \in L_2(S)$,

$$(3) \qquad \lim_{T \to \infty} (U_t f,g) = (f,1)(1,g)/\mu(S) .$$

If the flow is known to be ergodic, to prove strong mixing it is sufficient to verify (3) for those functions f with $(f,1) = 0$. In terms of the spectrum alone we see that φ_t is strongly mixing if it is ergodic and the spectrum of U_t, except for $\lambda = 0$, is absolutely continuous with respect to Lebesgue measure. The converse of this last condition, however, is not true. For the proofs of these statements (except the preceding sentence) see E. Hopf [1].

Now the equations (1) and (2) may be interpreted in another way. The mapping $t \to U_t$ is a representation of the additive group of the reals into the set of unitary operators on some Hilbert space. Since this group is abelian, its only irreducible unitary representations are in one-dimensional spaces, and (2) expresses the fact that the given representation is a combination of certain irreducible representations $t \to e^{i\lambda t}$. Questions about the spectrum of U_t may be rephrased as questions about which irreducible representations occur in U_t and what their multiplicities are. Let us review the pertinent representation theory and carry out these reformulations.

§3. Let G be a separable topological group. A homomorphism $U: g \to U_g$ of G into the group of all unitary transformations of a Hilbert space $\mathfrak{H}(U)$ is called a <u>continuous unitary representation</u> of G if $(U_g\varphi, \psi)$ is continuous for each φ, ψ in $\mathfrak{H}$. A unitary representation is <u>irreducible</u> if no proper closed linear manifold in $\mathfrak{H}$ is carried into itself by all the transformations U_g, g in G. If a subspace $\mathfrak{M}$ is invariant under all the U_g (we also say "$\mathfrak{M}$ reduces U") and E is the orthogonal projection of $\mathfrak{H}$ on $\mathfrak{M}$, then $V: g \to U_gE$ is a <u>subrepresentation</u> of U and denoted by $\mathfrak{M}_U$ or E_U. In this case V is said to be contained in U. Two unitary representations L and M of G are (unitarily) <u>equivalent</u> if there exists an isometry W of $\mathfrak{H}(L)$ onto $\mathfrak{H}(M)$ such that $WL_g = M_gW$ for all $g \in G$. If $\mathfrak{M}$ is the maximal closed linear manifold on which U_g is the identity operator for all g, we say the identity representation is contained in U with multiplicity equal to the dimension of $\mathfrak{M}$.

The decomposition theorem of Mautner will be used repeatedly in the sequel: If U is a continuous unitary representation of a separable, locally compact group G in a separable Hilbert space $\mathfrak{H}$, there exists a direct integral decomposition of $\mathfrak{H}$, $\int \oplus \mathfrak{H}^{\alpha} \, d\mu(\alpha)$, in which U is decomposable, $U = \{U^{\alpha}\}$, and for almost all α, U^{α} is an irreducible unitary representation on $\mathfrak{H}^{\alpha}$. (For definitions of the direct integral and proofs of this theorem, see Naimark [1] or Mackey [1], Chapter II.) Stone's theorem (2) may be considered as the special case of this general result obtained when G is the additive group of the reals.

Considering the map $U: t \to U_t$ of a flow as a unitary representation of the reals, we may restate the conditions for ergodicity and mixing. The flow is ergodic if and only if the identity representation is contained in U with multiplicity one. The flow is weakly mixing if it is ergodic and U has no finite dimensional (and hence, one-dimensional) subrepresentations other than the identity.

Actually, the situation we shall investigate is the following. The space S will be acted on (pointwise) transitively by a group G, and, in fact, will be a compact homogeneous space G/D, where G is a Lie group and D a closed subgroup. The flow φ_t is defined by restricting the natural action of G on S to a one-parameter subgroup, $\{g_t\}$. In all the cases we consider there will be a natural measure μ on G/D, which is invariant under the action of G. If $f \in L_2(G/D, \mu)$, the equation

$$(U_g f)(s) = f(s \cdot g)$$

defines a unitary representation of G. (Here, if s is the coset Dh, $s \cdot g$ is the coset Dhg.) Then $U_t = U_{g_t}$.

Suppose it has been possible to obtain some information about the direct integral decomposition of U as a representation of G. In particular, we may know the way U^α acts on $\mathfrak{h}^\alpha$ for every α. U^α restricted to $\{g_t\}$ will generally not be irreducible, but the analysis of its spectrum may be an easier task than that of U_t as a whole. First we notice that, by the transitivity of G on G/D, the identity representation is contained in U with multiplicity one (that is, the constants are the only functions invariant under all of G). Then we may let $\mathfrak{h}^{\alpha_0} = \mathbf{C}$ and $U^{\alpha_0} = I$. Suppose U_t^α has no non-zero invariant element for almost all $\alpha \neq \alpha_0$. Then the identity is contained in U_t also with multiplicity one, or the flow is ergodic. The same argument works for other eigenvalues. Suppose further, that in some direct integral decomposition $\mathfrak{M} = \int \oplus \mathfrak{M}^\alpha \, d\nu(\alpha)$ (where $\mathfrak{M}$ is possibly a subspace of $\mathfrak{h}$), U_t^α has absolutely continuous or Lebesgue spectrum for almost all α. (U_t has Lebesgue spectrum if $(E(\Delta)\varphi,\varphi)$ is equivalent to Lebesgue measure for every φ.) Then U_t has absolutely continuous or Lebesgue spectrum, respectively. The proof of this last statement follows quickly from the definitions of the spectral resolution and the direct integral, and may be found in detail in Fomin and Gelfand [1], §2.

§4. The most important method of constructing irreducible representations of topological groups is that of induced representations. This theory for non-compact groups is almost entirely due to Mackey.

Let H be a closed subgroup of the separable locally compact topological group G, and suppose $L\colon h \to L_h$ is a unitary representation of H in the Hilbert space $\mathfrak{h}(L)$. Let μ be a measure on the Borel subsets A of G/H which is quasi-invariant in the sense that $\mu(Ax) = 0$ if and only if $\mu(A) = 0$, where Ax denotes the image of A under the natural action of $x \in G$. If f,g are functions from G to $\mathfrak{h}(L)$ such that

(i) $$L_h f(x) = f(hx), \qquad h \in H,$$

and similarly for g, then the inner product $(f(x),g(x))$ in $\mathfrak{h}(L)$ is a function constant on right H cosets, so it may be considered a function on G/H. In particular, this is true of $\|f(x)\|$. $\mathfrak{h}$ is defined to be the totality of functions f satisfying (i) and two further conditions:

(ii) $(f(x),\varphi)$ is a Borel function of x for every $\varphi \in \mathfrak{h}(L)$,

(iii) $$\int_{G/H} \|f(x)\|^2 \, d\mu < \infty .$$

Let $\mu_x(A) = \mu(Ax)$ and put $\gamma_x(Hy) = \frac{d\mu_x}{d\mu}(Hy)$. Set $(U_x f)(y) = f(yx)\gamma_x^{1/2}(Hy)$. Then $x \to U_x$ is a unitary representation of G, called the representation U^L induced by L.

Because of (i), the functions in $\mathfrak{h} = \mathfrak{h}(U^L)$ are essentially determined by their values at one point from each coset of H. Let B be a Borel section of H in G; that is, a Borel subset of G which intersects each right coset Hg in exactly one point, $\xi(g)$. (Such sections exist: Mackey [2], Lemma 1.1.) Then $\beta(g) = g[\xi(g)]^{-1}$ is an H-valued Borel function on G. Setting $\mathfrak{B}(g) = L_{\beta(g)}$, we see that $\mathfrak{B}(hg) = L_h \mathfrak{B}(g)$ for $h \in H$ and $g \in G$. Let $L_2(G/H;\mathfrak{h}(L))$ be the Hilbert space of functions from G/H to $\mathfrak{h}(L)$ which are square-integrable with respect to the quasi-invariant measure μ. Then the equation

$$(V_x f)(Hy) = \mathfrak{B}^{-1}(y)\mathfrak{B}(yx)f(Hyx)\gamma_x^{1/2}(Hy) \tag{4}$$

can be easily seen to define a unitary representation of G. If $(\Gamma f)(g) = \mathfrak{B}(g)f(Hg)$, we compute that $V_x = \Gamma^{-1}U_x^L\Gamma$ for every $x \in G$, so V is equivalent to the induced representation U^L. In particular, the equivalence class of V is independent of the choice of B. This version of the induced representation often leads to simpler explicit formulas since the space G/H is smaller than G.

If L is the identity representation of H and μ is invariant $(\gamma_x \equiv 1)$, then (4) reads

$$(V_x f)(Hy) = f(Hyx) ;$$

hence the representation introduced in §3 is an induced representation.

When G is the semi-direct product of a normal H and another closed subgroup, that subgroup is a natural choice for $\mathfrak{B}$. This will be the case in the examples below and in Chapter V. In fact, in all these cases H is normal and $G/H \cong \mathbf{R}$, the additive group of the reals. Denoting a homomorphism of G into $\mathbf{R}$ by ρ, formula (4) becomes

$$(V_g f)(x) = \mathfrak{A}(g;x)f(x + \rho(g)) , \tag{5}$$

with $f \in L_2(\mathbf{R};\mathfrak{h}(L))$. If $g = h \cdot r$, $h \in H$, and $\rho(u) = x$, then $\mathfrak{A}(g;x) = L_{uhu^{-1}}$. Here Lebesgue measure on $\mathbf{R}$ is clearly invariant under the action of G.

In order to apply the method of induced representations, it is natural to ask for which groups is every irreducible representation equivalent to an induced representation, and, if we have such a group, which subgroups need be considered.† A rather general answer is given by Takenouchi [1]:

† Actually, most known representations of Lie groups are induced representations.

Theorem A. Let G be a connected solvable Lie group for which the exponential map from its algebra is surjective. Then every irreducible unitary representation is equivalent to a representation induced by a one-dimensional representation of a connected subgroup.

Every group whose representations we use in later chapters is of this kind; in particular, the nilpotent groups of Chapter V, the solvable subgroups of the simple group in Chapter III, and the solvable, non-rigid motion groups of Chapters III, and VII. What is more important for our present purpose, Takenouchi tells exactly how to go about finding all these representations. Call an irreducible representation of G completely irreducible if it is trivial on no connected normal subgroup N. (If L is not completely irreducible, it is essentially a representation of a smaller factor group.) Let L be a completely irreducible representation of a group G to which Takenouchi's theorem applies. Then the center, C, of G has dimension zero or one. There is a connected normal abelian subgroup H containing C such that C is of codimension one or two in H. If $\chi: h \to \chi(h)$ is a character of H (irreducible representation of dimension one), put $\chi^g(h) = \chi(ghg^{-1})$ for each $g \in G$. The set $\{\chi^g: g \in G\}$ is called the orbit of χ. The subgroup H_χ of all elements g such that $\chi^g = \chi$ is called the isotropy (or stabilizer) group of χ. H_χ is connected and clearly contains H. There is an extension $\tilde{\chi}$ of χ to a representation of H_χ, and every completely irreducible representation of G is of the form $U^{\tilde{\chi}}$ for suitable choice of $\tilde{\chi}$. To obtain a complete set of mutually inequivalent such representations, it is only necessary to take one χ from each orbit and then find all their inequivalent extensions $\tilde{\chi}$.

The proofs of these results involve direct applications of Mackey's general theory, and are found in detail in Takenouchi's paper [1]. In Chapter V, where nilpotent groups are discussed, we shall not form the representations in exactly this fashion. Since, however, the process is analogous and references are supplied for the detailed justification, we shall not go into the matter any further here. For the other applications we use only two specific groups, and we shall obtain their representations in the next section by Takenouchi's method.

§5. We consider two real solvable Lie algebras and their connected, simply-connected groups. S_1 has the basis $\{X,Y\}$ and relation $[X,Y] = Y$. Its associated group, S_1, is the affine group of the real line: $x \to ax + b$, $a > 0$.‡ The pairs $\{(a,b): a > 0\}$ multiply according to the rule $(a,b)(a',b') = (aa',ab' + b)$; in fact, the group may also be represented as the set of real matrices of the form $\begin{pmatrix} a & b \\ 0 & 1 \end{pmatrix}$ under ordinary matrix multiplication. Clearly the Lie algebra of S_1 is $\mathfrak{S}_1$, with

‡ The following description of the representations of S_1 is essentially that of Mackey [1], p. 169. These representations were first found by Gelfand and Naimark.

$\exp(cX) = (e^c,0)$ and $\exp(bY) = (1,b)$. The center of S_1 is trivial and the only connected normal abelian subgroup is $H = \{(1,b)\}$. In fact, since this is the only connected normal subgroup, the irreducible representations are either lifts of representations of the abelian group S_1/H, and hence one-dimensional, or they are induced by the process described in the preceding paragraph from characters of H. If $\chi_s\colon (1,b) \to e^{isb}$ is a character of H, $\chi_s^g(1,b') = \chi_s^{(a,b)}(1,b') = \chi_s(1,ab') = \chi_{as}(1,b')$. Thus there are three orbits: $\theta_0 = \{\chi_0\}$, $\theta_+ = \{\chi_s;\ s > 0\}$ and $\theta_- = \{\chi_s;\ s < 0\}$. $H_{\chi_0} = S_1$, so the extensions of χ_0 are the lifts of characters of S_1/H. $H_{\chi_1} = \{(a,b)\colon \chi_1^{(a,b)} = \chi_1\}$ is clearly H itself, as is $H_{\chi_{-1}}$. Therefore, proper extensions of χ_1 and χ_{-1} need not be considered, and the only other representations are those induced by these two, call them U^+ and U^-, respectively. Now S_1 is the semi-direct product of H and $K = \{(a,0)\colon a > 0\}$. A homomorphism of S_1 onto the additive reals is given by $\rho((a,b)) = \log a$. According to the description given in (5) of the previous section, we can then write down explicitly what these induced representations are:

$$(U^{\pm}_{(a,b)}f)(x) = \exp(\pm\, ie^x b)f(x + \log a), \qquad f \in L_2(\mathbf{R}) \tag{6}$$

Theorem B. (Gelfand-Naĭmark) Every irreducible unitary representation of S_1 is equivalent to either U^+ or U^- given by formula (6), or is a one-dimensional representation $U^\beta(a,b) = \exp(i\beta \log a)$, β real.

Corollary. (Mautner's Lemma) Let $U\colon g \to U_g$ be a unitary representation of S_1 in a Hilbert space $\mathfrak{h}$. Let $\mathfrak{M}$ be the closed linear manifold of all $\psi \in \mathfrak{h}$ such that $U_h\psi = \psi$ for all $h \in H$. Then the spectrum of $U(\exp tX)$ restricted to $\mathfrak{M}^\perp$ is Lebesgue; in particular, every eigenvector of $U(\exp tX)$ is in $\mathfrak{M}$.

Proof. Apply Mautner's decomposition theorem (§3) to the representation U. $\mathfrak{h} = \int \oplus\, \mathfrak{h}^\alpha d\mu(\alpha)$. Let A_0 be the set of α such that $\mathfrak{h}^\alpha$ is one-dimensional, A_1 its complement. $\mathfrak{h}_0$ and $\mathfrak{h}_1$ will denote the corresponding direct integrals over these sets. Since H is the commutator of S_1, $\mathfrak{h}_0 \subseteq \mathfrak{M}$. For $0 \neq \psi \in \mathfrak{h}_1$, $h = (1,b) \in H$, and $\alpha \in A_1$,

$$(U_h^\alpha \psi^\alpha)(x) = \exp(\pm\, ie^x b)\psi^\alpha(x),$$

so $\psi \notin \mathfrak{M}$. Thus $\mathfrak{M} = \mathfrak{h}_0$. Now for $\alpha \in A_1$,

$$(U^\alpha(\exp tX)\psi^\alpha)x = \psi^\alpha(x + t),$$

which is the familiar translation operator in $L_2(\mathbf{R})$. This one-parameter group has Lebesgue spectrum, so by the result quoted at the end of §3, $U(\exp tX)$ has Lebesgue spectrum on $\mathfrak{h}_1 = \mathfrak{M}^\perp$.

We now derive analogous results for the group S_2 whose Lie algebra $\mathfrak{S}_2$ has a basis $\{X,Y,Z\}$ and relations $[X,Z] = \sigma Y + Z$, $[X,Y] = Y - \sigma Z$, $[Y,Z] = 0$, where σ is a non-zero real number. S_2 may be represented by pairs $\{(t,w)\}$ with t real and w complex. Here $\exp(tX) = (t,0)$, $\exp(aY + bZ) = (0,a + ib)$, and the group multiplication is

$$(t,w)(t',w') = (t + t', e^{(1-i\sigma)t} w' + w) .$$

S_2 is the semi-direct product of the one-parameter group $\{(t,0)\}$ and the two-dimensional normal abelian subgroup $H = \{(0,w)\}$. The center is trivial, and, being the only abelian normal subgroup, H must be the group to use in finding the completely irreducible representations.

For each complex number z we obtain a character χ_z of H by the formula $\chi_z((0,w)) = \exp[i\Re(zw)]$. Let $g = (t',w')$. Then we calculate that $\chi_z^g = \chi_{z'}$, where $z' = e^{(1-i\sigma)t'} z$. Thus each orbit, for $z \neq 0$, is a spiral in the z-plane. Moreover, the isotropy group χ_z is H for $z \neq 0$, while for $z = 0$ it is, of course, all of S_2. We are again in the situation of the previous example, where proper extensions of the characters are not needed (except for χ_0). However, there is now a continuum of distinct orbits. A convenient cross-section of them is the set of χ_r, where r is a real number between 1 and $e^{2\pi/\sigma}$. The homomorphism ρ is of course $\rho((t,w)) = t$, so the representation induced by χ_r, which we denote by U^r, has the form

$$(7) \qquad (U^r_{(t,w)}f)(x) = \exp[ir\Re(e^{(1-i\sigma)x} w)]f(x + t), \qquad f \in L_2(\mathbf{R})$$

<u>Theorem C</u>. Every irreducible unitary representation of S_2 is equivalent to a U^r given by formula (7), with r in the half-open interval whose end points are 1 and $e^{2\pi/\sigma}$, or is a one-dimensional representation $U^\beta_{(t,w)} = \exp(i\beta t)$, β real.

<u>Corollary D</u>. Let U be a unitary representation of S_2 in a Hilbert space. Let $\mathfrak{M}$ be the largest manifold on which U restricted to H is the identity. Then $U(\exp tX)$ restricted to $\mathfrak{M}^\perp$ has Lebesgue spectrum; every eigenvector of $U(\exp tX)$ is in $\mathfrak{M}$.

The proof of Corollary D is almost word for word the same as the proof of Mautner's Lemma. We omit the details.

In conclusion, we remark that for either $\mathfrak{S}_1$ or $\mathfrak{S}_2$ the Lie subalgebra corresponding to H may be defined without reference to bases as

$$\mathfrak{S}^\infty = \lim_{k \to \infty} \underbrace{[\mathfrak{S},[\mathfrak{S},\ldots,[\mathfrak{S}}_{k},\mathfrak{S}]\ldots]] .$$

For $\mathfrak{S}_1$ and $\mathfrak{S}_2$, of course, the sequence is constant and $\mathfrak{S}^\infty = [\mathfrak{S},\mathfrak{S}]$. In Chapter VII we shall see that, with this definition of H, Corollary D holds for all simply-connected solvable groups for which the exponential map is surjective. Since $\mathfrak{S}^\infty$ reduces to zero if G is nilpotent, other methods are needed to analyze the nilflows.

CHAPTER III.

FLOWS ON SOME THREE DIMENSIONAL HOMOGENEOUS SPACES

by

L. Auslander
L. Green
F. Hahn

§1. Introduction. This chapter concerns itself with three main problems: (1) to determine all the connected, simply connected, non-compact, three-dimensional Lie groups G which have a discrete subgroup D such that G/D is compact; (2) for each such group G determine all the discrete subgroups D such that G/D is compact; (3) for the groups in (2) give a description of the dynamical behavior of the flows induced on the compact spaces G/D by the one-parameter subgroups of G. These examples are of interest not only because of their exhaustive nature but also because they exhibit different dynamical behaviors which are liable to occur in higher dimensions.

We are interested in the following characteristics of the flows: are there dense orbits? are all the orbits dense? is the flow ergodic? if it is ergodic is it mixing? Each one of our spaces G/D has a measure which is invariant under the one-parameter flows and does not vanish on open sets. If such a flow is ergodic [Gottschalk-Hedlund, Th, 9.20] we see that there are no invariant open sets and thus there exist dense orbits. The flow is called minimal if all the orbits are dense. This is a topological porperty which cannot be derived from ergodicity or the spectral properties of the flow as discussed in Chapter II. Even in dimension three, we have minimal flows with pure point spectrum (irrational flows on a torus) mixed spectrum (nilflows) and continuous spectrum (generalized horocycle flows).

Problems (1) and (2) are answered in section two of this chapter. For completeness section three begins with a brief discussion of the geodesic and horocycle flows of Hopf and Hedlund. This section then, closes with the solution to problem (3) the results of which are sketched in the following table. (Note: the theorems in the text are usually more detailed than the table.)

Group	Flow	Dynamical Character
I. Simple	1) Periodic	Periodic.
	2) Generalized Horocycle	Minimal, ergodic, strongly mixing.
	3) Generalized Geodesic	Non-minimal, ergodic, strongly mixing.
II. Solvable but not nilpotent		
a) S_1	1) Type 1	No dense orbit.
	2) Type 2	Non-minimal, ergodic, mixed spectrum.
b) S_2	1) Type 1	No dense orbit.
	2) Type 2	Periodic and/or torus flow depending on the subgroup D.
III. Nilpotent		
a) non-abelian		Determined completely by associated torus flow.
b) abelian		Torus-flow.

§2. The Groups. In this section we are interested in classifying all three-dimensional connected, simply connected, non-compact Lie groups G which have discrete subgroups D such that G/D is compact. We will also classify all the discrete subgroups D for each G such that G/D is compact. Since for each three-dimensional Lie algebra there is a unique connected, simply-connected Lie group, and vice versa, a first attempt at this problem is to list all the three-dimensional Lie algebras. If X, Y, Z is a basis for the Lie algebra then the following multiplication tables describe all the non-isomorphic Lie algebras of dimension 3 (for details see Jacobson).

2.1. Abelian. $[X,Y] = [X,Z] = [Y,Z] = 0$

2.2. Nilpotent. $[X,Y] = Z, \quad [X,Z] = [Y,Z] = 0$

2.3. Solvable. $[X,Y] = 0, \quad [X,Z] = \alpha X + \beta Y$

$$[Y,Z] = \gamma X + \delta Y, \quad \text{where}$$

$$A = \begin{pmatrix} \alpha & \beta \\ \gamma & \delta \end{pmatrix} \quad \text{is non singular.}$$

2.4. Simple. There are only two distinct simple algebras and a basis may be chosen such that

a) $[X,Y] = -2Y, \quad [X,Z] = 2Z, \quad [Y,Z] = X,$

or

b) $[X,Y] = Z, \quad [X,Z] = -Y, \quad [Y,Z] = X.$

In the abelian case the corresponding Lie group G is always isomorphic to the group of matrices

$$\begin{pmatrix} 1 & 0 & 0 & x_1 \\ 0 & 1 & 0 & x_2 \\ 0 & 0 & 1 & x_3 \\ 0 & 0 & 0 & 1 \end{pmatrix}$$

where the x_i are real numbers and under the isomorphism D is mapped onto the subgroup

$$\begin{pmatrix} 1 & 0 & 0 & n_1 \\ 0 & 1 & 0 & n_2 \\ 0 & 0 & 1 & n_3 \\ 0 & 0 & 0 & 1 \end{pmatrix}$$

where n_i are integers.

In the nilpotent case the corresponding Lie group G is always isomorphic to the matrix group

$$\begin{pmatrix} 0 & x_1 & x_2 \\ 0 & 1 & x_3 \\ 0 & 0 & 1 \end{pmatrix}$$

where the x_i are real numbers and D is mapped onto a subgroup generated by the matrices

$$\begin{pmatrix} 1 & 1 & 0 \\ 0 & 1 & 0 \\ 0 & 0 & 1 \end{pmatrix}, \begin{pmatrix} 1 & 0 & 0 \\ 0 & 1 & 1 \\ 0 & 0 & 1 \end{pmatrix}, \begin{pmatrix} 1 & 0 & \frac{1}{k} \\ 0 & 1 & 0 \\ 0 & 0 & 1 \end{pmatrix},$$

where k is a fixed positive integer. (For details see Chapter IV.)

We need only examine in detail the solvable and simple cases. In the solvable case we will show that there are only two non-isomorphic connected, simply-connected, non-compact, solvable non-nilpotent Lie groups which we call S_1 and S_2. The group S_1 is isomorphic to the matrix group

$$\begin{pmatrix} e^{kz} & 0 & 0 & x \\ 0 & e^{-kz} & 0 & y \\ 0 & 0 & 1 & z \\ 0 & 0 & 0 & 1 \end{pmatrix}$$

where x, y, and z are real numbers and k is a fixed real number such that $e^k + e^{-k}$ is an integer different from 2. In this case the matrix A of 2.3 is given by $\begin{pmatrix} -k & 0 \\ 0 & k \end{pmatrix}$.

The group S_2 is isomorphic to the matrix group

$$\begin{pmatrix} \cos 2\pi z & \sin 2\pi z & 0 & x \\ -\sin 2\pi z & \cos 2\pi z & 0 & y \\ 0 & 0 & 1 & z \\ 0 & 0 & 0 & 1 \end{pmatrix}$$

where x, y, z are real numbers. In this case the matrix A of 2.3 is given by $\begin{pmatrix} 0 & 2\pi \\ -2\pi & 0 \end{pmatrix}$. In the next section we will classify the discrete subgroups D of S_i such that S_i/D is compact.

In the simple case the Lie algebra of 2.4a is isomorphic to the Lie algebra of the Lie group of two by two real matrices of determinant 1. We call this group $G(2)$. This group is not simply connected. The unique connected, simply-connected, non-compact, simple Lie group G which corresponds to the Lie algebra in 2.4a has no matrix representation. However, there is a homomorphism $\eta\colon G \to G(2)$ whose kernel is a discrete central subgroup of G and η is a covering map. In section (4) we will examine the discrete subgroups D of G such that G/D is compact and show that $\eta(D)$ is discrete in $G(2)$ and that G/D is a finite covering of $G(2)/\eta(D)$.

The unique connected, simply-connected Lie group which corresponds to the Lie algebra in 2.4b is the three-dimensional spinor group. But this group is a two fold covering of the three-dimensional real orthogonal group and is thus compact.

§3. We now examine the solvable case in detail.

LEMMA 3.1. If S is a connected, simply-connected, non-compact, non-nilpotent, three-dimensional Lie group with maximal nilpotent subgroup N and a discrete subgroup D such that S/D is compact, then the dimension of N is two.

PROOF. The dimension of N is not zero since if it were the commutator of S would be trivial and S would be abelian which is not so. Since S is not nilpotent we see that dimension $N \neq 3$. We must only show that dimension $N \neq 1$ and we do this by showing that the assumption dimension $N = 1$ leads to a contradiction.

If dimension $N = 1$ then the commutator subgroup R of S has dimension one or zero. Since S is not abelian R must have dimension one and is consequently isomorphic to the reals. Since R is a normal subgroup of S we see that S acts on R by letting each $s \in S$ act as follows $r \to s^{-1}rs$. From Mostow's work [1], we see that $D \cap R$ is infinite cyclic. Since R is normal and D is a subgroup, we see that $D \cap R$ is invariant under the action of D. There are only two automorphisms of an infinite cyclic group so there is a subgroup D' of D of index two such that D' acts trivially on $D \cap R$ and thus acts trivially on R. If H is the normal subgroup of S which leaves R pointwise fixed, we see that $H \supset D'$.

Now S/D is compact and D/D' has order two so S/D' is compact and consequently S/H is compact. We obtain an action of S/H on R by letting each $sH \in S/H$ act on R as follows: $r \to s^{-1}rs$. Since S/H is compact and connected it follows that this action is trivial (Montgomery-Zippin [1]) and thus R is in the center of S. This implies that S is nilpotent which is the needed contradiction.

With the aid of this theorem we can complete the announced classification of the solvable groups with discrete subgroups whose quotient is compact.

THEOREM 3.2. If S is a connected, simply-connected, non-compact, non-nilpotent, three-dimensional Lie group with a discrete subgroup D such that S/D is compact then S is isomorphic to either S_1 or S_2.

PROOF. In order to prevent the technical details of the proof from obscuring the line of thought we sketch the theorem. We will first show that $S = R \cdot R^2$, where R is the additive group of reals and the dot means semi-direct product. Consequently S is completely determined once we know how R acts on R^2. If η: $S/R^2 = R$ is the natural map we observe that $\eta(R^2 \cap D)$ is a non-trivial closed subgroup of R generated by a single element θ. R is thus isomorphic to a one-parameter subgroup of $G\ell(R,2)$ which goes through θ. This subgroup will be identified and then the theorem will be complete.

If we let N be the maximal nilpotent subgroup of S the previous lemma tells us that $N = R^2$. The group S satisfies the exact sequence

$$1 \to R^2 \to S \to R \to 1$$

where $R = S/N$. Since N is abelian the factor group S/N acts on N in the following manner: if $sN \in S/N$ then its action on $n \in N$ is given by $n \to sns^{-1}$. By Iwasawa, we have $S = R \cdot R^2$.

From Mostow's work [1] it follows that $D \cap N$ is a lattice generated by two linearly independent elements. We choose coordinates (x,y) in R^2 such that $D \cap N$ is generated by $(1,0)$ and $(0,1)$. The compactness of S/D shows that D is not contained in N and its projection on S/N is a closed non-trivial subgroup of R generated by a single element θ. We observe that θ acts on R^2 as a linear transformation which preserves the lattice generated by $(0,1)$ and $(1,0)$.

At this point we should observe that we have completely described the subgroup D. For if $\theta \cdot (u,v)$ is a preimage of θ in D then $\theta \cdot (u,v)$, $(0,1)$ and $(1,0)$ generate D. Since θ is non-singular, we may choose $(u',v') \in R^2$ such that $(\theta - I)(u',v') = (u,v)$. The inner automorphism of $R \cdot R^2$ induced by (u',v') leaves N pointwise fixed and sends $\theta \cdot (u,v) \to \theta$. The image of D under this automorphism is generated by θ, $(0,1)$, and $(1,0)$.

We must still determine how R acts on R^2. This is equivalent to finding the one-parameter subgroups $\varphi(t)$ of $G\ell(R,2)$ for which $\varphi(1) = \theta$. Since θ preserves the integral lattice of R^2 its matrix with respect to $(1,0)$ and $(0,1)$ has integral entries and has determinant ± 1. θ cannot have a negative determinant since it lies on a one-parameter subgroup $\varphi(t)$ of $G\ell(R,2)$. Consequently $\varphi(t)$ is a one-parameter subgroup of the unimodular group. There are three distinct cases to consider. First the eigenvalues of θ are real and positive. Second, the eigenvalues of θ are real and negative. Third, the eigenvalues are complex with non-trivial imaginary parts.

In the first case if we write θ in diagonal form, we see that a square root of it must be in diagonal form. In fact we can easily compute that the only 2×2 matrix which is in diagonal form, with determinant one, and has a square root not in diagonal form must have trace -2. It is thus the matrix with -1 on the diagonal. θ must thus have two square roots, one with positive eigenvalues and one with negative eigenvalues not -1. The second alternative cannot be since such a transformation has no further square root and thus cannot lie on a one-parameter subgroup. By taking successive square roots we see that there is a unique one-parameter subgroup $\varphi(t)$ such that $\varphi(1) = \theta$. This one-parameter subgroup is isomorphic to the group

$$\left\{ \begin{pmatrix} e^{kz} & 0 \\ 0 & e^{-kz} \end{pmatrix} : \ z \in R, \quad e^k + e^{-k} \text{ a positive integer} \right\} .$$

In the second case our previous argument on square roots shows that the eigenvalues of θ are -1. In this case θ lies on a compact one-parameter subgroup which is unique. R thus acts on R^2 as a continuous group of rotations.

In the third case the eigenvalues of θ are complex conjugates with product one and sum an integer. The only possible pairs of such eigenvalues are:

$$e^{\pi i} \text{ and } e^{-\pi i}, \quad e^{\frac{\pi i}{2}} \text{ and } e^{\frac{-\pi i}{2}}, \quad e^{\frac{\pi i}{3}} \text{ and } e^{\frac{-\pi i}{3}}, \quad e^{\frac{2\pi i}{3}} \text{ and } e^{\frac{-2\pi i}{3}} .$$

In each of these cases θ lies on a unique compact one-parameter subgroup and R acts on R^2 as a group of rotations.

We easily see that $S = R \cdot R^2$ is isomorphic to S_1 or S_2 where the isomorphism is given by

$$z \cdot (x,y) \to \begin{pmatrix} \multicolumn{2}{c}{A(z)} & 0 & x \\ & & 0 & y \\ 0 & 0 & 1 & z \\ 0 & 0 & 0 & 1 \end{pmatrix} .$$

COROLLARY 3.3. The only discrete subgroups D of S_1 such that S_1/D is compact are generated by

$$\begin{pmatrix} e^k & 0 & 0 & 0 \\ 0 & e^{-k} & 0 & 0 \\ 0 & 0 & 1 & 1 \\ 0 & 0 & 0 & 1 \end{pmatrix}, \quad \begin{pmatrix} 1 & 0 & 0 & u_1 \\ 0 & 1 & 0 & v_1 \\ 0 & 0 & 1 & 0 \\ 0 & 0 & 0 & 1 \end{pmatrix}, \quad \begin{pmatrix} 1 & 0 & 0 & u_2 \\ 0 & 1 & 0 & v_2 \\ 0 & 0 & 1 & 0 \\ 0 & 0 & 0 & 1 \end{pmatrix}$$

where

$$\begin{vmatrix} u_1 & v_1 \\ u_2 & v_2 \end{vmatrix} \neq 0.$$

COROLLARY 2.8. If D is a discrete subgroup of S_2 such that S_2/D is compact then either D is generated by

$$\begin{pmatrix} \cos 2\pi n/p & \sin 2\pi n/p & 0 & 0 \\ -\sin 2\pi n/p & \cos 2\pi n/p & 0 & 0 \\ 0 & 0 & 1 & n/p \\ 0 & 0 & 0 & 1 \end{pmatrix}$$

$$\begin{pmatrix} 1 & 0 & 0 & u_1 \\ 0 & 1 & 0 & v_1 \\ 0 & 0 & 1 & 0 \\ 0 & 0 & 0 & 1 \end{pmatrix} \quad \begin{pmatrix} 1 & 0 & 0 & u_2 \\ 0 & 1 & 0 & v_2 \\ 0 & 0 & 1 & 0 \\ 0 & 0 & 0 & 1 \end{pmatrix}$$

where n is an integer, p is $2, 3, 4$ or 6 and $\begin{vmatrix} u_1 & v_1 \\ u_2 & v_2 \end{vmatrix} \neq 0$;

or D is generated by

$$\begin{pmatrix} 1 & 0 & 0 & 1 \\ 0 & 1 & 0 & 0 \\ 0 & 0 & 1 & 0 \\ 0 & 0 & 0 & 1 \end{pmatrix} \quad \begin{pmatrix} 1 & 0 & 0 & 0 \\ 0 & 1 & 0 & 1 \\ 0 & 0 & 1 & 0 \\ 0 & 0 & 0 & 1 \end{pmatrix} \quad \begin{pmatrix} 1 & 0 & 0 & u \\ 0 & 1 & 0 & v \\ 0 & 0 & 1 & n \\ 0 & 0 & 0 & 1 \end{pmatrix}$$

where n is an integer.

§4. We now let $G(2)$ be the group of two by two real matrices of determinant one and let G be its universal covering group and η the covering homomorphism. G is the unique connected, simply-connected, non-compact, simple, three-dimensional Lie group. If D is a discrete subgroup of G, such that G/D is compact, we will show that $\eta(D)$ is a discrete subset of $G(2)$ and G/D is a finite covering of $G(2)/\eta(D)$.

LEMMA 4.1. If A is an abelian subgroup of $G(2)$ containing elements not in the center of $G(2)$ and if D is a subgroup of $G(2)$ which is in the normalizer of A, then at least one of the following possibilities occurs.

a) D contains a completely reducible subgroup H normal in D and of index at most two in D.

b) D is reducible.

c) D is a subgroup of a rotation group.

PROOF. We distinguish three cases: either there is an element $a' \in A$ which has real distinct eigenvalues; or there are no elements of A with real distinct eigenvalues but there is a non-central element $a' \in A$ with real equal eigenvalues; or all the non-central elements of A have complex eigenvalues.

In the first case we will first show that A is completely reducible. Let $\lambda_1 \neq \lambda_2$ be real eigenvalues of a' and let v_1, v_2 be the corresponding eigenvectors. Since A is abelian, we have for each $a \in A$ $a'a(v_i) = aa'(v_i) = \lambda_i a(v_i)$, $i = 1,2$. Thus, $a(v_i)$ is an eigenvector of a' with eigenvalue λ_i and consequently $a(v_i)$ is a multiple of v_i by some numbers $\mu_i(a)$ depending on a, that is $a(v_i) = \mu_i(a) \cdot v_i$. Therefore v_i are distinct eigenvectors for each $a \in A$ so A is completely reducible.

Since A is normal in D we see that for any $d \in D$ there is an $a \in A$ such that $a'd = da$. Consequently we have, $a'(d(v_i)) = d(a(v_i)) = \mu_i(a) \cdot d(v_i)$ and therefore $d(v_i)$ is an eigenvector of a'. Thus the elements of D can at most permute the two eigenspaces. The set of elements H of D which leave the eigenspaces invariant, is a normal subgroup of D of index at most two and we have the alternative a) of the conclusion of the theorem.

We now suppose that we have case two and $a' \in A$ is non-central with the single real eigenvalue λ and an eigenvector v. We show that v is the only eigenvector for the non-central elements of A. Since the determinant of a' is one, we see that $\lambda = \pm 1$ and since a' is non-central, v is the only eigenvector of a'. The same argument as before shows that v is an eigenvector for all of A and thus only eigenvector for the non-central elements of A.

Again an argument similar to that used in the first case shows that D leaves the eigenspace $\{v\}$ invariant and thus D is reducible, which is alternative b).

We assume now that the third case holds and we show first that A is contained in a rotation group. Arguing as before, we see that A is completely reducible over the complex field. Thus, there is a two by two complex matrix b such that $bab^{-1} = \begin{pmatrix} e^{iy} & 0 \\ 0 & e^{-iy} \end{pmatrix}$ for each $a \in A$, and y is a real number depending on a. We let $D' = bDb^{-1}$. Since A must be normal in D', we see by a straightforward matrix computation that the elements of D' are either of the form $d_1' = \begin{pmatrix} e^{ix} & 0 \\ 0 & e^{-ix} \end{pmatrix}$, x real or $d_2' = \begin{pmatrix} 0 & w \\ \frac{1}{w} & 0 \end{pmatrix}$, w complex $\neq 0$. We know that $b^{-1}d_1'b$ is real and then we compute that $b^{-1}d_2'b$ cannot be real. Since $b^{-1}D'b = D$ we see that there are no elements of the type d_2' in D' and thus D is contained in the same circle group containing A. This completes the theorem.

THEOREM 4.2. Let η be the projection of G onto $G(2)$, and let K be the kernel of η. If D is a discrete subgroup of G such that G/D is compact, then $K \cdot D/D$ is finite.

PROOF. We will show that if $K \cdot D/D$ is infinite then $G(2)/\overline{\eta(D)}$ is not compact. It would then follow that $G/\eta^{-1}(\overline{\eta D})$ is a covering of $G(2)/\overline{\eta(D)}$. But there is a continuous map of G/D onto $G/\eta^{-1}(\overline{\eta D})$. Hence G/D would not be compact. We show first that if $K \cdot D/D$ is infinite then $K \cdot D$ is not discrete and thus the identity component H of $\overline{K \cdot D}$ is not zero dimensional. We thus show that H is abelian. Once this is done, we complete the proof as follows: $\eta(H)$ is a non-central abelian subgroup of $G(2)$ and $\eta(D)$ is in the normalizer of $\eta(H)$ and thus is one of the three types mentioned in the previous lemma. In any one of these cases $G(2)/\overline{\eta(D)}$ is not compact.

We now need only prove the assertions made about KD. If $K \cdot D/D$ is infinite, it follows from the compactness of G/D that there is a $g \in G$ and a sequence k_i in K, in distinct cosets of D, such that $k_iD \to gD$ in G/D. Thus, there exist d_i in D such that $k_id_i \to g$. Now G is connected and K is a normal discrete subgroup so it must be central. This together with the convergence of k_id_i implies that $d_id_{i+1}^{-1}k_ik_{i+1}^{-1} \to e$ and consequently KD is not discrete.

We now let H be the identity component of $\overline{KD}$ and choose a neighborhood U of e such that U meets D in only the identity and such that $g,g' \in U$ implies $[g,g'] \in U$. If $k,k' \in K$ and $d,d' \in D$ such that $kd,k'd' \in U$ then $[d,d'] = [kd,k'd'] \in U$. This means that $[d,d'] = e$ and thus the commutators in $U \cap H$ are trivial so H is abelian. This completes the proof.

COROLLARY 4.3. If D is a discrete subgroup of G such that G/D is compact, then $gD \to \eta(g)\eta(D)$ defines a finite covering of $G(2)/\eta(D)$ by G/D and $\eta(D)$ is discrete in $G(2)$.

PROOF. We need only observe that the preimage of $\eta(D)$ under the covering map is $K \cdot D/D$. By the previous theorem, this is finite and since D is discrete, we see $K \cdot D$ is discrete and consequently so is $\eta(D)$.

We complete this section by describing those discrete subgroups D of G for which G/D is compact. We begin with the following known lemma whose proof is included for the sake of completeness.

LEMMA 4.4. Let Z be an infinite cyclic group with generator z and let B be a group generated by $b_1,\ldots,b_n$ with a single defining relation $w = b_{i_1}^{\alpha_1} \ldots b_{i_j}^{\alpha_j} = e$. If A satisfies

$$1 \to Z \to A \xrightarrow{\varphi} B \to 1$$

and Z is in the center of A, then A is generated by $n+1$ elements $a_1,\ldots,a_n,z$ and has $n+1$ defining relations

$$vz^{\alpha} = a_{i_1}^{\alpha_1} \ldots a_{i_j}^{\alpha_j} z^{\alpha} = e$$

$$a_iz = za_i \qquad i = 1,\ldots,n \quad .$$

PROOF. We choose a_i such that $\varphi(a_i) = b_i$ and we assert that $a_1,\ldots,a_n,z$ generate A. If $a \in A$ let $\varphi(a) = b$. This is a word in the b_i so there is a word $\bar{a}$ in the a_i for which $\varphi(\bar{a}) = b$. We see that $\varphi(a^{-1}\bar{a}) = b^{-1}b = e$ so $a^{-1}\bar{a}$ is in Z the kernel of φ. We can therefore write $a = \bar{a}z^{\beta}$ which proves our assertion.

We now need to show that any other relation is a consequence of the n+1 relations listed above. We suppose $a = a_{j_1}^{\beta_1} \ldots a_{j_n}^{\beta_n} z^{\beta} = e$. Since $\varphi(a) = e$ in B and this group has only one defining relation $w = e$ we see that $\varphi(a) = \prod_{i=1}^{m} t_i w^{\delta_i} t_i^{-1}$ where the t_i are words in the b_i. It follows that $a = \left(\prod_{i=1}^{m} s_i v^{\delta_i} s_i^{-1}\right) \cdot z^{\gamma}$, where the s_i are preimages of the t_i and v is defined in the theorem. We thus have $e = a = \left(\prod_{i=1}^{m} s_i (vz^{\alpha})^{\delta_i} s_i^{-1}\right) z^{\gamma-\alpha\varepsilon}$ where $\varepsilon = \sum_{i=1}^{m} \delta_i$. Since z generates a free subgroup of A it follows that $\gamma-\alpha\varepsilon = 0$ and thus $a = e$ is a consequence of the n+1 relations listed.

We next give the structure of the fundamental group of the tangent sphere bundle to a compact orientable surface of constant negative curvature. To do this easily we make the following conventions. We let $G^* = G(2)/Z_2$ where $Z_2 = \left\{\begin{pmatrix} 1 & 0 \\ 0 & 1 \end{pmatrix}, \begin{pmatrix} -1 & 0 \\ 0 & -1 \end{pmatrix}\right\}$ and define $\eta^*\colon G \to G^*$ by $g \to g^* = \eta(g)Z_2$. We let R be the unique one-parameter subgroup which contains the kernel of η and we let the circle $C^* = \eta^*(R)$. We let D^* be a discrete subgroup of G^* such that G^*/D^* is compact. We have the following diagram:

$$\begin{array}{ccccccc} 1 \to R & \to & G & \longrightarrow & R\backslash G & \longrightarrow & 1 \\ \downarrow \eta^* & & \downarrow \eta^* & & \downarrow \eta^* & & \\ C^* & \to & G^*/D^* & \to & C^*\backslash G^*/D^* & & \end{array}$$

For simplicity all mappings downward are called η^* and are defined in the obvious manner from the map η^*. It is known [for more detail see the beginning of the next section] that the double coset space $C^*\backslash G^*/D^*$ is a compact 2-dimensional orientable manifold M of constant negative curvature, and G^*/D^* is its tangent sphere bundle T, and the coset mapping is the fiber mapping of T onto M. Conversely, any such manifold M along with its sphere bundle T determines a diagram as above.

THEOREM 4.5. Let M be a compact orientable surface of constant negative curvature with Euler-Poincaré characteristic N and genus $h = \frac{1}{2}(N + 2)$. Let T be the tangent sphere bundle of M. If $\Pi(\ )$ is the fundamental group of whatever occurs inside the parenthesis then $\pi(T)$ is generated by $2h+1$ generators $a_1, a_2, \ldots, a_h$, $b_1, \ldots, b_h, z$ with the following defining relations

$$a_1 b_1 a_1^{-1} b_1^{-1} \cdot \ldots \cdot a_h b_h a_h^{-1} b_h^{-1} z^N = e$$

$$a_i z = z a_i$$

$$b_i z = z b_i \qquad i = 1, \ldots, h$$

PROOF. We take $D^* \subset G^*$ so that $M = C^* \backslash G^* / D^*$. Our previous diagram then becomes

$$\begin{array}{ccccccc}
1 \to & R & \to & G & \longrightarrow & R\backslash G & \longrightarrow 1 \\
\eta^* & \downarrow & & \downarrow \eta^* & & \downarrow \eta^* & \\
& S & \to & T & \longrightarrow & M & \\
& \| & & \| & & \| & \\
& C^* & & G^*/D^* & & C^*\backslash G^*/D^* &
\end{array}$$

In each column the inverses of the coset determined by $e \in G^*$ is the group of deck transformations of R, G, and $R\backslash G$ respectively. Since these are simply connected covering spaces of S, T, and M we see that $\pi(S) \subset R$ and $\pi(T) \subset G$ (Seifert and Threlfall [1]) and in particular $\pi(S)$ is in the kernel of η so it is central in $\pi(T)$. The bottom horizontal row induces the exact sequence

$$1 \longrightarrow \pi(S) \to \pi(T) \to \pi(M) \to 1$$

Since M is an orientable surface of genus h we know (Seifert and Threlfall [1]) that $2h = N + 2$ and that $\pi(M)$ has $2h$ generators $\bar{a}_1, \ldots, \bar{a}_h$ $\bar{b}_1, \ldots, \bar{b}_h$ with one defining relation

$$\bar{a}_1 \bar{b}_1 \bar{a}_1^{-1} \bar{b}_1^{-1} \cdot \ldots \cdot \bar{a}_h \bar{b}_h \bar{a}_h^{-1} \bar{b}_h^{-1} = 1$$

We now apply Lemma 4.4. In order to see that in the defining relations of $\pi(M)$ the power of z is N, we use Theorem 3 proved by Massey in the appendix of this chapter. See also Seifert [1].

Let $D \subset G$ be a discrete subgroup such that G/D is compact and let $D^* = \eta(D)$. Call $M = C^* \backslash G^* / D^*$.

THEOREM 4.6. If D is a discrete subgroup of G such that G/D is compact and if M is a manifold then D is generated by $2h + 1 = N + 3$ generators $a_1, \ldots, a_h$, $b_1, \ldots, b_h, z$ with the $2h + 1$ defining relations

$$wz^p = a_1 b_1 a_1^{-1} b_1^{-1} \ldots a_h b_h a_h^{-1} b_h^{-1} z^p = e$$

$$a_i z = z a_i$$

$$b_i z = z b_i \qquad i = 1 \ldots h$$

and p divides N the Euler-Poincaré characteristic of M.

PROOF. We recall the diagram of the previous theorem and the discussion of the fundamental groups of S, T, and M. We see that $D \subset \pi(T)$ and it maps onto $\pi(M) = R\backslash R \cup D$. We therefore see that D satisfies a sequence

$$1 \to Z \to D \to \pi(M) \to 1$$

where $Z \subset \pi(S)$ and $\pi(S)$ is generated by z_o, Z is generated by $z = z_o^n$. If we pick the generators a_i, b_i, $i = 1 \dots h$ as before, we see that a_i, b_i, z_o $(i = 1,\dots,h)$ generate $\pi(T)$ while a_i, b_i, z $(i = 1,\dots,h)$ generate D. Since by Theorem 4.5 $wz_o^N = e$, we see that $z_o^N = w^{-1} \in D$. Thus $z_o^N \in Z \subset \pi(S)$. Hence p divides N and the theorem is completed by applying Lemma 4.4.

COROLLARY 4.7. Let D be a discrete subgroup of G such that G/D is compact. Then there is a discrete subgroup D* of G* such that $\eta^* D = D^*$ and the kernel K of η^* is non-trivial and in the center of G. Further, if D_1^* is any subgroup of D* of finite index in D* such that $M = C^*\backslash G^*/D_1^*$ is a manifold, then the generator of K divides the Euler-Poincaré characteristic of M.

§5. <u>The Flows</u>. Before we begin to prove the theorems listed in the table in section 1 we recall briefly the description of the geodesic and horocycle flows on a surface of constant negative curvature.

We let M be the two dimensional Riemannian manifold whose space is the set of complex numbers z with positive imaginary part and whose differential metric is given by

$$ds^2 = \frac{dx^2 + dy^2}{y^2} \qquad \text{where} \quad z = x + iy .$$

We let U be the bundle of unit tangents vectors of M. If v is a unit tangent vector at z then there is a unique geodesic through z with v as its tangent. The geodesic flow on U is given as follows: each unit tangent v moves along the geodesic it determines in the direction of v with unit velocity.

A horocycle in M is either a circle tangent to the real axis or a straight line parallel to the real axis. If v is a unit tangent at z which does not point vertically away from the real axis then there is a unique horocycle such that v is an inward normal to the horocycle. If v points vertically away from the real axis then the horocycle determined by v is the line parallel to the real axis and through the base point of v. To each unit tangent v at z we may associate a unique unit tangent v' at z so that v',v have the same orientation as the positive real and positive complex axes respectively. The horocycle flow on U is described as follows: the unit tangent v at z moves with unit velocity along the horocycle it determines in the direction of v'.

The group Σ of linear fractional transformations which maps the upper half plane onto itself acts simply transitively on U and is thus homeomorphic to U. We let v_0 be the unit tangent at i whose direction is the same as the imaginary axis and always consider the homeomorphism determined by $\sigma \to \sigma(i)$ where $\sigma \in \Sigma$. Each element of Σ may be written $\sigma(z) = \frac{az + b}{cz + d}$ where a,b,c,d are real and ad - bc = 1. If we let G(2) be the group of two by two real matrices of determinant one there is a natural continuous mapping of $G(2) \to \Sigma$ given by $\begin{pmatrix} a & b \\ c & d \end{pmatrix} \to \frac{az + b}{cz + d} = \sigma(z)$. The kernel H of this mapping is the two subgroup $Z_2 = \left\{\begin{pmatrix} 1 & 0 \\ 0 & 1 \end{pmatrix}, \begin{pmatrix} -1 & 0 \\ 0 & -1 \end{pmatrix}\right\}$ of G(2). We thus see that U is homeomorphic to $G(2)/Z_2$.

If we trace the movement of the initial element v_0 under the geodesic flow and then use the transitivity of Σ on U, we see that the geodesic flow on U is isomorphic to the flow on $G(2)/Z_2$ induced by the one-parameter subgroup $\varphi(t) = \begin{pmatrix} e^t & 0 \\ 0 & e^{-t} \end{pmatrix}$. In the same manner we see that the horocycle flow is isomorphic to the flow in $G(2)/Z_2$ induced by the one-parameter subgroup $\varphi(t) = \begin{pmatrix} 1 & t \\ 0 & 1 \end{pmatrix}$.

A compact surface of constant negative curvature is covered by the hyperbolic plane M and is the orbit space M/Σ', where Σ' is a properly discontinuous subgroup of Σ. Thus the geodesic and horocycle flows on compact surfaces of constant negative curvature may always be realized on $G(2)/Z_2D$, where D is a discrete subgroup of G(2) for which $G(2)/D$ is compact and the flow is induced by $\begin{pmatrix} e^t & 0 \\ 0 & e^{-t} \end{pmatrix}$ or $\begin{pmatrix} 1 & t \\ 0 & 1 \end{pmatrix}$, respectively.

We will use Corollary 4.3 and the following theorem to take advantage of known facts about the geodesic and horocycle flows.

THEOREM 5.1. Let (X,T) and (Y,T) be transformation groups, with X and Y compact connected n-dimensional manifolds. Let θ: $X \to Y$ be a local homeomorphism onto Y such that $\theta(x_t) = (\theta(x))_t$. Under these conditions (X,T) is minimal if and only if (Y,T) is minimal.

PROOF. We recall first that minimal means that each orbit is dense. This of course is equivalent to saying there are no closed proper invariant subsets. We observe that compactness of X implies that θ is a closed mapping and thus for each $A \subset X$ we have $\theta(\bar{A}) = \overline{\theta(A)}$.

If (X,T) is minimal and $y \in Y$ then there is an $x \in X$ such that $\theta(x) = y$. Consequently $y_T = (\theta(x))_T = \theta(x_T)$ and thus $\bar{y}_T = \theta(\bar{x}_T) = \theta(X) = Y$, so (Y,T) is minimal.

If (X,T) is not minimal then there is a point in X whose orbit is not X. The orbit closure of this point must contain a minimal set M (Gottschalk-Hedlund [1]) which is not X. We observe that M contains no interior points. For, if M did have an interior point, then each point of M would be interior and M would be open as well as closed. The connectivity of X would then say M = X which is not the case. Since θ is a local homeomorphism $\theta(M)$ has no interior and is closed. Thus,

$\theta(M) \neq Y$ and is invariant so Y is not minimal.

The previous theorem will allow us to lift the minimality property of the horocycle flow. It seems more difficult to lift ergodicity and mixing. However, Mautner's proof [1] that the geodesic flow is mixing applies word for word to the generalized geodesic flows. The generalized horocycle flows are then treated by reversing the steps E. Hopf [2] used to prove mixing of the geodesic flow.

LEMMA 5.2. (Mautner) Let G be any connected Lie group which has the Lie algebra 2.4a, and let U be a continuous unitary representation of G in a Hilbert space $\mathfrak{H}$. Let $\mathfrak{H}_0$ be the largest subspace on which U is the identity. Then $U\ (\exp tX)$ has absolutely continuous spectrum on $\mathfrak{H}_0^{\perp}$.

PROOF. Let G_1 and G_2 be the subgroups of G corresponding to the subalgebras generated by $\{X,Y\}$ and $\{X,Z\}$, respectively. Each of these is a homomorphic image of the solvable, simply-connected group S of affine motions of the real line. Hence U restricted to G_i is a unitary representation of S, $i = 1,2$, and the corollary to Theorem B of Chapter II applies. Let $\mathfrak{M}_1, \mathfrak{M}_2$ be the closed linear manifolds of $\mathfrak{H}$ consisting of all vectors invariant under $U\ (\exp tY)$, $U\ (\exp tZ)$, respectively. By the corollary, $U\ (\exp tX)$ has Lebesgue spectrum on $\mathfrak{M}_i^{\perp}$, $i = 1,2$. Let $\varphi \in \mathfrak{M}_1^{\perp} + \mathfrak{M}_2^{\perp}$, and write $\varphi = \varphi_1 + \varphi_2$, $\varphi_1 \in \mathfrak{M}_i^{\perp}$. If Δ is a Borel subset of the line with Lebesgue measure zero, $\|E(\Delta)\varphi_i\| = 0$ for $i = 1,2$, where $E(\cdot)$ is the spectral resolution for $U\ (\exp tX)$. Then $\|E(\Delta)\varphi\| \leq \|E(\Delta)\varphi_1\| + \|E(\Delta)\varphi_2\| = 0$, or the spectrum of $U\ (\exp tX)$ is absolutely continuous on $M_1^{\perp} + M_2^{\perp}$. Now

$$(\mathfrak{M}_1^{\perp} + \mathfrak{M}_2^{\perp})^{\perp} \subseteq \mathfrak{M}_1 \cap \mathfrak{M}_2 = \mathfrak{H}_0 ,$$

since the subgroups $\{\exp tY\}$ and $\{\exp tZ\}$ generate G. Thus $\mathfrak{H}_0^{\perp} \subseteq \mathfrak{M}_1^{\perp} + \mathfrak{M}_2^{\perp}$, and the lemma is proved.

We now wish to examine the one-parameter flows on the homogeneous spaces of simple Lie groups. We describe briefly the groups and algebras involved. We let $\mathfrak{G}$ be the Lie algebra of two by two matrices of trace zero. G will be the unique simply-connected Lie group with this algebra and $G(2)$ will be the group of two by two matrices with determinant one. If $Z_2 = \left\{\begin{pmatrix}1 & 0\\ 0 & 1\end{pmatrix}, \begin{pmatrix}-1 & 0\\ 0 & -1\end{pmatrix}\right\}$ we get $G^* = G(2)/Z_2$. Each of these groups, G, $G(2)$, G^*, has the Lie algebra $\mathfrak{G}$ and we have the following commutative diagram

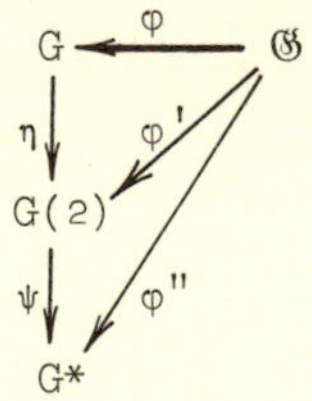

where φ, φ' and φ'' are the appropriate exponential maps and η and ψ are the appropriate covering maps.

The geodesic and horocycle flows take place on G^* modulo, a discrete subgroup D^*, such that G^*/D^* is compact. If D is a discrete subgroup of G we will study the one-parameter flows on G/D by relating them to the known geodesic and horocycle flows on G^*/D^*, where $D^* = \psi \cdot \eta(D)$ The fact that D^* is discrete follows from Corollary 4.3. We define a covering map θ: $G/D \to G^*/D^*$ given by $\theta(gD) = \psi\eta(g) \cdot D^* = g^*D^*$. This covering map θ has the following important properties: it defines a finite covering of G^*/D^* (see Corollary 4.3); if $x \in \mathfrak{G}$ then $\theta(\varphi(X)gD) = \varphi''(X)g^*D^*$. With these properties in mind we prove:

THEOREM 5.3. Let G be the connected, simply-connected, non-compact, three-dimensional Lie group. Let D be a discrete subgroup of G for which G/D is compact. Let X be in the Lie algebra of G. One of the following statements is valid.

a) (Generalized geodesic flow) If X has real distinct eigenvalues, then the flow $(G/D, \varphi(Xt))$ is ergodic, strongly mixing and has infinitely many closed orbits.

b) (Generalized horocycle flow) If X has real equal eigenvalues, then the flow $(G/D, \varphi(Xt))$ is ergodic, strongly mixing and minimal.

c) (Periodic flow) If X has complex eigenvalues, then the flow $(G/D, \varphi(Xt))$ is periodic.

PROOF. a) We suppose that X has real distinct eigenvalues. There is an inner automorphism of $\mathfrak{G}$ which diagonalizes X. This induces an inner automorphism on G and G^* which commutes with the appropriate exponential maps. In view of this, we may assume, without loss of generality, that X is in diagonal form $\begin{pmatrix} x & 0 \\ 0 & -x \end{pmatrix}$. We then have that $\varphi'(Xt) = \begin{pmatrix} e^{xt} & 0 \\ 0 & e^{-xt} \end{pmatrix}$ and $(G^*/D^*, \varphi''(Xt))$ is the geodesic flow. The geodesic flow is known to be ergodic, strongly mixing and have infinitely many closed orbits (Hedlund [2] and Hopf [1][2]). The map θ: $G/D \to G^*/D^*$, described before, maps the flow $(G/D, \varphi(Xt))$ onto the flow $(G^*/D^*, \varphi''(Xt))$. Since this is a finite covering of the geodesic flow it follows that $(G/D, \varphi(Xt))$ has infinitely many closed orbits.

Let U: $g \to U(g)$ be the continuous unitary representation of G in $L_2(G/D)$ obtained by the natural action of G on the homogeneous space. Since G acts transitively on G/D, the fixed point space of U is the one-dimensional space spanned by the constant functions, $\mathfrak{H}_0$. By Lemma 5.2, $U(\varphi(Xt))$ has absolutely continuous spectrum on $\mathfrak{H}_0^{\perp}$. In particular, the only functions invariant under $U(\varphi(Xt))$ are constant, so the flow is ergodic. Absolute continuity of the spectrum implies strong mixing, as noted in Chapter II, §2.

b) If X has equal real eigenvalues we may, as before, assume that X is in the form $\begin{pmatrix} 0 & x \\ 0 & 0 \end{pmatrix}$. In this case $\varphi'(Xt) = \begin{pmatrix} 1 & xt \\ 0 & 1 \end{pmatrix}$ and $(G/D, \varphi(Xt))$ finitely covers the horocycle flow $(G^*/D^*, \varphi''(Xt))$. The horocycle flow is minimal (Hedlund [1]) and it follows from Theorem 5.1

that $(G/D, \varphi(Xt))$ is minimal.

The proof that $(G/D, \varphi(Xt))$ is ergodic is more involved than in the previous case. The group-theoretic calculation of the spectrum of the horocycle flow (Parasyuk, [1]) was made in the spirit of Gelfand-Fomin [1] using the specific forms of all the irreducible unitary representations of $G(2)$. Since such a description does not seem to be presently available for the covering groups of $G(2)$, and as Mautner's method does not apply to this flow, we will adapt Hopf's [2] original argument to the present situation.

Choose coordinates X, Y, Z in $\mathfrak{G}$ so that

$$X = \begin{pmatrix} 0 & 1 \\ 0 & 0 \end{pmatrix}, \quad Y = \begin{pmatrix} 0 & 0 \\ 1 & 0 \end{pmatrix}, \quad Z = \begin{pmatrix} 1 & 0 \\ 0 & -1 \end{pmatrix} .$$

Hopf has established the following relation in $G(2)$ by geometric reasoning:

$$(1) \qquad \exp\left[\frac{\alpha(t)}{2}(X-Y)\right] \exp\left(\frac{s(t)}{2} Z\right) \exp\left[\frac{\alpha(t)-\pi}{2}(X-Y)\right] = \exp(-tX) .$$

Here

$$(2) \qquad s(t) = \frac{1}{2}\log\left(\frac{\sqrt{4+t^2}+t}{2}\right), \quad \alpha(t) = \arctan(2/t) .$$

(The geometric meaning of (1) is clarified by constructing the geodesic of length $s(t)$ joining the ends of a horocycle arc of length t in the hyperbolic plane.) Equation (1) may be verified in $G(2)$ by direct calculation. In G itself, the same relation must hold modulo elements of the center, K. Because K is discrete and (1) holds in the neighborhood of $t = 0$, (1) actually holds for all t in G. Notice, also, that as $t \to \infty$, we have $\alpha(t) \to 0$ and $s(t) \to \infty$.

Let gD denote a point of G/D, $d\mu$ the measure obtained from Haar measure. For $f \in L_2(G/D,\mu)$, define the unitary operators

$$(D_\gamma f)(gD) = f(\exp[\tfrac{\gamma}{2}(X-Y)]gD)$$

$$(T_s f)(gD) = f(\exp[\tfrac{s}{2}Z]gD)$$

$$(H_t f)(gD) = f(\exp[tX]gD) .$$

Then (1) may be rewritten

$$H_{-t}f = D_{\alpha-\pi} T_s D_\alpha f ,$$

when s and α are given by equations (2). Now suppose that $(f,1) = 0$ and g is arbitrary in $L_2(G/D,\mu)$.

$$|(H_{-t}f,g)| = |(D_{\alpha-\pi}T_sD_\alpha f,g)| = |(T_sD_\alpha f,D_{\pi-\alpha}g)|$$

$$\leq |(T_sD_\alpha f,D_{\pi-\alpha}g) - (T_sD_\alpha f,D_\pi g)|$$

$$+ |(T_sD_\alpha f,D_\pi g) - (T_sf,D_\pi g)| + |(T_sf,D_\pi g)|$$

$$\leq \|T_s D_\alpha f\| \; \|D_{-\alpha} D_\pi g - D_\pi g\| + \|T_s D_\alpha f - T_s f\| \; \|D_\pi g\| + |(T_s f, D_\pi g)|$$

$$= \|f\| \; \|D_{-\alpha} D_\pi g - D_\pi g\| + \|D_\alpha f - f\| \; \|D_\pi g\| + |(T_s f, D_\pi g)| \; .$$

As $t \to \infty$, the terms involving norms approach zero because D_α approaches the identity strongly as $\alpha \to 0$. The last term approaches zero because the geodesic flow is strongly mixing. This completes the proof that the flow determined by X is strongly mixing and ergodic.

c) If X has complex eigenvalues then since its trace is zero its eigenvalues must be of the form $-ix$, ix. The matrix $\begin{pmatrix} 0 & x \\ -x & 0 \end{pmatrix}$ has these eigenvalues. Consequently X is similar to this matrix and we may, as before, assume $X = \begin{pmatrix} 0 & x \\ -x & 0 \end{pmatrix}$. In this case $\varphi'(Xt) = \begin{pmatrix} \cos xt & \sin xt \\ -\sin xt & \cos xt \end{pmatrix}$. The flow $(G^*/D^*, \varphi''(Xt))$ is thus periodic. Since $(G/D, \varphi(Xt))$ is a finite covering of this periodic flow it follows that the flow $(G/D, \varphi(Xt))$ is periodic.

§6. We will now examine the flows on the solvmanifolds of section 3. The groups S_1 and S_2 will be those three-dimensional matrix groups described in section 3. The Lie algebras $\mathfrak{S}_i$ of these groups are the matrix algebras given by

$$\begin{pmatrix} \multicolumn{2}{c}{A_i(c)} & 0 & a \\ & & 0 & b \\ 0 & 0 & 0 & c \\ 0 & 0 & 0 & 0 \end{pmatrix} ,$$

where a,b,c are real numbers and $A_1(c) = \begin{pmatrix} kc & 0 \\ 0 & -kc \end{pmatrix}$, $A_2(c) = \begin{pmatrix} 0 & 2\pi c \\ -2\pi c & 0 \end{pmatrix}$.

For the remainder of this section we will fix the following bases for $\mathfrak{S}_i$, $i = 1,2$

$$X = \begin{pmatrix} 0 & 0 & 0 & 1 \\ 0 & 0 & 0 & 0 \\ 0 & 0 & 0 & 0 \\ 0 & 0 & 0 & 0 \end{pmatrix}, \quad Y = \begin{pmatrix} 0 & 0 & 0 & 0 \\ 0 & 0 & 0 & 1 \\ 0 & 0 & 0 & 0 \\ 0 & 0 & 0 & 0 \end{pmatrix},$$

$$Z = \begin{pmatrix} \multicolumn{2}{c}{A_i(1)} & 0 & 0 \\ & & 0 & 0 \\ 0 & 0 & 0 & 1 \\ 0 & 0 & 0 & 0 \end{pmatrix} .$$

For the sake of brevity, we introduce the following notation. Let $B_i(z) = \exp A_i(z)$ for $i = 1,2$. We then let

$$\begin{pmatrix} x \\ y \\ z \end{pmatrix} = \begin{pmatrix} \multicolumn{2}{c}{B_i(z)} & 0 & x \\ & & 0 & y \\ 0 & 0 & 1 & z \\ 0 & 0 & 0 & 1 \end{pmatrix}$$

The choice of $i = 1,2$ depends, of course, on the group we are using.

In either group S_i, we distinguish two types of one-parameter subgroups.

Type one: $\varphi(t) = \exp(aX + bY)t;$

Type two: $\varphi(t) = \exp(aX + bY + cZ)t, \quad c \neq 0.$

The following theorem is now immediate

THEOREM 6.1. Let D be a discrete subgroup of S_i such that S_i/D is compact $(i = 1,2)$. If $\varphi(t)$ is a one-parameter group of type one in S_i then the flow $(S_i/D, \varphi(t))$ $(i = 1,2)$ has no dense orbits.

We must now examine the flows induced by type two one-parameter subgroups of the groups S_i.

THEOREM 6.2. Let D be a discrete subgroup of S_1 such that S_1/D is compact. Let $\varphi(t)$ be a one-parameter subgroup of type two. Then the flow $(S_1/D, \varphi(t))$ has a periodic orbit and is thus not minimal.

PROOF. It follows from Corollary 3.3 that there is an automorphism of S_1 such that the image of D is generated by

$$\begin{pmatrix} 0 \\ 0 \\ 1 \end{pmatrix}, \begin{pmatrix} u_1 \\ u_2 \\ 0 \end{pmatrix}, \begin{pmatrix} v_1 \\ v_2 \\ 0 \end{pmatrix}, \quad \text{where} \quad \begin{vmatrix} u_1 & v_1 \\ u_2 & v_2 \end{vmatrix} \neq 0.$$ Without loss of generality we may assume that D is this subgroup.

Since $\varphi(t)$ is of type two in S_1 we compute

$$\varphi(t) = \begin{matrix} a(e^{kct} - 1) \\ b(e^{-kct} - 1) \\ ct \end{matrix} \quad , \quad c \neq 0.$$

We examine the orbit of the point $\begin{pmatrix} -a \\ -b \\ 0 \end{pmatrix} D \in S_1/D$ and find its position at $t = \frac{1}{c}$.

$$\begin{pmatrix} -a \\ -b \\ 0 \end{pmatrix} D_{\frac{1}{c}} = \varphi(\tfrac{1}{c}) \begin{pmatrix} -a \\ -b \\ 0 \end{pmatrix} D = \begin{pmatrix} -a \\ -b \\ 1 \end{pmatrix} D$$

$$= \begin{pmatrix} -a \\ -b \\ 0 \end{pmatrix}\begin{pmatrix} 0 \\ 0 \\ 1 \end{pmatrix} D = \begin{pmatrix} -a \\ -b \\ 0 \end{pmatrix} D .$$

Thus we see that this point lies on a periodic orbit and the theorem is complete.

THEOREM 6.3. Let D be a discrete subgroup of S_2 such that S_2/D is compact. Let D be generated by three elements of the following type:

$$\begin{matrix} \cos 2\pi\, n/p & \sin 2\pi\, n/p & 0 & 0 & u_1 & v_1 \\ -\sin 2\pi\, n/p & \cos 2\pi\, n/p & 0 & 0 & u_2 & v_2 \\ 0 & 0 & 1 & n/p & 0 & 0 \\ 0 & 0 & 0 & 1 & & \end{matrix}$$

where n is an integer and $p = 2,3,4$ or 6, and $\begin{vmatrix} u_1 & v_1 \\ u_2 & v_2 \end{vmatrix} \neq 0.$

If $\varphi(t)$ is a one-parameter subgroup of type two in S_2, then the flow $(S_2/D, \varphi(t))$ is periodic.

PROOF. Since $\varphi(t)$ is of type two

$$\varphi(t) = \begin{pmatrix} a \sin 2\pi ct - b\,[\cos 2\pi\, ct - 1] \\ b \sin 2\pi ct + a\,[\cos 2\pi\, ct - 1] \\ ct \end{pmatrix}, \quad c \neq 0 .$$

If m is an integer then $\varphi(\frac{pm}{c}) = \begin{pmatrix} 0 \\ 0 \\ pm \end{pmatrix}$ and $\varphi(\frac{pm}{c})$ is in the center of S_2 and belongs to D. The following computations then show that the flow $(S_2/D, \varphi(t))$ has period $\frac{pm}{c}$:

$$\begin{pmatrix} x \\ y \\ z \end{pmatrix} D_{\frac{pm}{c}} = \varphi(\tfrac{pm}{c}) \begin{pmatrix} x \\ y \\ z \end{pmatrix} D = \begin{pmatrix} x \\ y \\ z \end{pmatrix} \varphi(\tfrac{pm}{c})\, D = \begin{pmatrix} x \\ y \\ z \end{pmatrix} D .$$

We need the following lemma to complete the topological considerations on flows on these solvmanifolds.

LEMMA 6.4. Let $\varphi(t)$ be a one-parameter subgroup of type two in S_i. Then there is an inner automorphism of S_i for which the image of $\varphi(t)$ is of the form $\begin{pmatrix} 0 \\ 0 \\ ct \end{pmatrix} = \exp\,(cZ)t$.

PROOF. Since $\varphi(t)$ is of type two, it is of the form

$$\varphi(t) = \exp\,(aX + bY + cZ)t$$

where $Z = \begin{pmatrix} A_i(c) & & 0 & 0 \\ & & 0 & 0 \\ 0 & 0 & 0 & c \\ 0 & 0 & 0 & 0 \end{pmatrix}$. Since $A_i(c)$ is non-singular, there exist a_i, b_i such that $A_i(c) \begin{pmatrix} a_i \\ b_i \end{pmatrix} = \begin{pmatrix} a \\ b \end{pmatrix}$ for $i = 1,2$. If we let

$B = \begin{pmatrix} 1 & 0 & 0 & a_i \\ 0 & 1 & 0 & b_i \\ 0 & 0 & 1 & 0 \\ 0 & 0 & 0 & 1 \end{pmatrix}$ then $B(aX + bY + cZ)t\,B^{-1} = cZt$. Since $B(\exp C)B^{-1} = \exp BCB^{-1}$ for each $C \in \mathfrak{S}_i$ the lemma is complete.

THEOREM 6.5. Let L be the lattice of integer points in R^3. Let D be a discrete subgroup of S_2 for which S_2/D is compact and D is generated by three elements $\begin{pmatrix} 1 \\ 0 \\ 0 \end{pmatrix}\begin{pmatrix} 0 \\ 1 \\ 0 \end{pmatrix}\begin{pmatrix} u \\ v \\ n \end{pmatrix}$, n an integer. If $\varphi(t)$ is a one-parameter subgroup of type two in S_2 then the flow $(S_2/D, \varphi(t))$ is isomorphic to a straight line flow of slope $(u,v,-1)$ on R^3/L.

PROOF. Since the flow $(S_2/D, \varphi(t))$ is isomorphic to the flow $(D\backslash S_2, \varphi(t))$, we may prove the theorem for the flow on the right coset space $D\backslash S_2$. In the proof of the previous lemma, we saw that the inner automorphism induced by B leaves the subgroup D pointwise fixed. We may therefore assume, without loss of generality, that $\varphi(t) = \begin{pmatrix} 0 \\ 0 \\ ct \end{pmatrix}$, $c \neq 0$.

We let θ be the one-parameter subgroup of R^3 given by $\theta(t) = (-uct, -vct, ct)$. We will now define an isomorphism

$$\psi\colon\ (D\backslash S_2,\ \varphi(t)) \to (R^3/L,\ \theta(t))$$

by $\psi D\begin{pmatrix} x \\ y \\ x \end{pmatrix} = (x - uz,\ y - vz,\ z)L$. It can be verified that ψ is well defined and is a homeomorphism of $D\backslash S_2$ and R^3/L. We also see that

$$\left(\psi D\begin{pmatrix} x \\ y \\ z \end{pmatrix}\right)_t = (x - uz - uct,\ y - vt - vct,\ z + ct)D = \psi\ D\begin{pmatrix} x \\ y \\ z \end{pmatrix}_t .$$

Consequently ψ is the isomorphism sought and the theorem is complete.

We remark that flow of the type mentioned in Theorem 6.5 are now completely classified. The flow is minimal and ergodic if u, v, and 1 are rationally independent and is periodic otherwise (Koksma [1]).

THEOREM 6.7. Let D be a discrete subgroup of S_1 such that S_1/D is compact. If $\varphi(t)$ is a subgroup of type two in S_1 then the flow $(S_1/D, \varphi(t))$ is ergodic and has a mixed spectrum. Consequently, dense orbits exist for this flow. Hence the flow is neither mixing nor equi-continuous.

PROOF. It follows from Corollary 3.7 that we may assume that D is generated by three elements $\begin{pmatrix} 0 \\ 0 \\ 1 \end{pmatrix}, \begin{pmatrix} u_1 \\ u_2 \\ 0 \end{pmatrix}, \begin{pmatrix} v_1 \\ v_2 \\ 0 \end{pmatrix}$ where $\begin{vmatrix} u_1 & v_1 \\ u_2 & v_2 \end{vmatrix} \neq 0$. Since $\varphi(t)$ is of type two it must be of the form $\exp\,(aX + bY + cZ)t$, $c \neq 0$. By Lemma 6.4 there exists an inner automorphism leaving D pointwise fixed and mapping $\varphi(t)$ to $\exp cZt$. It is not difficult to see that a change in time scale does not affect the measure theoretic properties of a flow. Because of these considerations we may, without loss of generality, assume $\varphi(t) = \exp Zt$.

Let $[S_1, S_1]$ be the commutator of S_1 and let $\eta\colon S_1/D \to S_1/D[S_1, S_1]$ be the natural map. The map η is a homomorphism of the flow $(S_1/D, \varphi(t))$ onto the flow $(S_1/D[S_1,S_1], \varphi(t))$. The space of this second flow is a circle and $\varphi(t)$ acts on it by a rotation through $2\pi t$. We let $\mathfrak{H}_1 = L_2(S_1/D)$ and $\mathfrak{H}_2 = L_2(S_1/D[S_1,S_1])$. Let $U_i(g)$, $g \in S_1$, $(i = 1, 2)$ be the unitary representations with spaces $\mathfrak{H}_i$ $(i = 1, 2)$ induced by the transformation groups $(S_1/D, S_1)$ and $(S_1/D[S_1,S_1], S_1)$.

We now wish to show that there is a one to one correspondence, preserving eigenvalues, between eigenvectors of $U_1(\varphi(t))$ and $U_2(\varphi(t))$. The subgroups of S_1 with Lie algebras spanned by $\{X, Z\}$ and $\{Y, Z\}$ are isomorphic to the group to which Mautner's Lemma (corollary to Theorem B, Chapter II) applies. Thus every eigenvector of $U_1(\exp Zt)$ is left fixed by $U_1(\exp Xt)$ and $U_1(\exp Yt)$, for all t. If $x \in \mathfrak{H}_1$ is such an eigenvector, then $U_1(g)x = x$ for all $g \in [S_1, S_1]$. This implies that there is a unique $y \in \mathfrak{H}_2$ such that $x = \bar{\eta}(y)$, where $\bar{\eta}\colon \mathfrak{H}_2 \to \mathfrak{H}_1$ is the map induced by η: $S_1/D \to S_1/D[S_1,S_1]$. We thus see that $U_2(\exp Zt)y = e^{i\lambda t}y$ for all $t \in T$ if $U_1(\exp Zt)x = e^{i\lambda t}x$.

Conversely, if $y \in \mathfrak{H}_2$ and $U_2(\exp Zt)y = e^{i\lambda t}y$ for all $t \in T$, then $U_1(\exp Zt)x = e^{i\lambda t}x$ for all $t \in T$, where $x = \bar{\eta}(y)$. This yields the existence of the one to one correspondence between eigenvectors.

The eigenspace of $\mathfrak{H}_2$, $U_2(\varphi(t))$, belonging to the eigenvalue zero consists of the constant functions only and is thus one-dimensional. It follows that the eigenspace of $\mathfrak{H}_1$ belonging to the eigenvalue zero is one-dimensional and thus the flow $(S_1/D,\varphi(t))$ is ergodic. There are eigenvectors of $\mathfrak{H}_2$, $U_2(\varphi(t))$, with non zero eigenvalues. Hence such eigenvectors also exist for $\mathfrak{H}_1, U_1(\varphi(t))$. This shows that the spectrum of the flow $(S_1/D,\varphi(t))$ is discontinuous and thus the flow is not mixing. These eigenvectors do not span $\mathfrak{H}_1 = L_2(S_1/D)$ so the spectrum of $(S_1/D,\varphi(t))$ is mixed.

The fact that the flow $(S_1/D,\varphi(t))$ is ergodic shows that there is a dense orbit. If the flow were also equicontinuous, $(S_1/D,\varphi(t))$ would have to be minimal which contradicts Theorem 6.2. The flow is thus not equicontinuous and the proof of the theorem is complete.

Let S be a simply-connected, connected, non-compact, non-nilpotent, solvable, 3 dimensional Lie group with a discrete subgroup D such that S/D is compact. It follows as a corollary to the theorems of section 3 that the theorems of this section describe the behaviour of all the flows $(S/D, \varphi(t))$.

§7. The details for flows on nilmanifolds are worked out in the following two chapters. In this section we will merely state the results as they apply to three-dimensional nilmanifolds. The only non-abelian connected, simply-connected, non-compact, three-dimensional, nilpotent Lie group is isomorphic to the group N of three by three matrices with zeros below the diagonal, ones on the diagonal, and arbitrary real entries above the diagonal. The discrete subgroups D for which N/D is compact are isomorphic to subgroups generated by the elements.

$$(7.1)\qquad \begin{pmatrix} 1 & 1 & 0 \\ 0 & 1 & 0 \\ 0 & 0 & 1 \end{pmatrix}, \begin{pmatrix} 1 & 0 & 0 \\ 0 & 1 & 1 \\ 0 & 0 & 1 \end{pmatrix}, \begin{pmatrix} 1 & 0 & \frac{1}{k} \\ 0 & 1 & 0 \\ 0 & 0 & 1 \end{pmatrix},$$

where k is a fixed integer. This isomorphism may be extended to an automorphism of N (Malcev [1]). Thus, without loss of generality, we may always assume that D is generated by three elements as in (7.1). The one-parameter subgroups of N are of the form

$$\varphi(t) = \begin{pmatrix} 1 & at & ct + \frac{1}{2}abt^2 \\ 0 & 1 & bt \\ 0 & 0 & 1 \end{pmatrix}$$

We remark that a flow on a metric space is called distal if for each pair of points $p \neq q$ we have $\inf_{-\infty < t < +\infty} d(p_t,q_t) > 0$.

THEOREM 7.2. Let $D \subset N$ be a discrete subgroup such that N/D is compact. If $\varphi(t)$ is a one-parameter subgroup of N, then the flow $(N/D, \varphi(t))$ is always distal. The flow $(N/D, \varphi(t))$ is both minimal and ergodic if a and b are rationally independent. The flow is neither minimal nor ergodic if a and b are rationally dependent.

If A is a connected, simply-connected, abelian, three-dimensional Lie group, then A is isomorphic to R^3. Arguments similar to those used before show that if D is a discrete subgroup such that R^3/D is compact, then, without loss of generality, we may assume that D is the integral lattice of R^3. We thus only need the classical Kronecker Theorem on diophantine approximations (Koksma [1]) to obtain:

THEOREM 7.3. Let $\varphi(t) = (at, bt, ct)$ be a one-parameter subgroup of R^3. The flow $(R^3/D, \varphi(t))$ is minimal and ergodic if a, b, c are rationally independent and is neither minimal nor ergodic if a, b, c are rationally dependent.

CHAPTER III.

Appendix

ON THE FUNDAMENTAL GROUP OF CERTAIN FIBRE SPACES.

by

W. Massey

§1. Introduction. Let G be a connected topological group which is an Eilenberg-MacLane Space of type $K(\pi,1)$ with π abelian and let B be a connected CW-complex which is an Eilenberg-MacLane Space of type $K(\Pi,1)$. Given any principal G-bundle $p : E \longrightarrow B$ over B, we can ask, what is the fundamental group of E? The only non-trivial part of the homotopy sequence of this bundle is the following:

$$0 \longrightarrow \pi_1(G) \longrightarrow \pi_1(E) \longrightarrow \pi_1(B) \longrightarrow 0$$

This shows that $\pi_1(E)$ is a group extension over $\pi_1(B) = \Pi$ with kernel $\Pi_1(G) = \pi$. Corresponding to this extension there is a certain 2-dimensional cohomology class $k \in H^2(\Pi,1;\pi)$ which determines the equivalence class of the extension completely. On the other hand, the characteristic class of this fibre bundle (i.e., the first obstruction to a cross section) is a 2-dimensional cohomology class $c \in H^2(B,\pi_1(G)) = H^2(\Pi,1;\pi)$. It is natural to conjecture that these two cohomology classes are the same. In this note, we will essentially prove this conjecture, modulo automorphisms of $H^2(\Pi,1;\pi)$, for the case where Π acts trivially on π.

As an application of this result, one can determine the fundamental group of the bundle of unit tangent vectors to a compact, orientable 2-manifold of genus ≥ 0.

§2. Statement and Proof of the Theorems. Let G be an Eilenberg-MacLane Space $K(\pi,1)$ with π a countable abelian group. Following Milnor [1], we will take G to be an abelian topological group. We will consider equivalence classes of principal G-bundles over a fixed base space B which is assumed to be a CW-complex. There are two ways of introducing an addition in this set of equivalence classes of bundles:

(a) The classifying space B_G is an Eilenberg space $K(\pi,2)$, where π is the fundamental group of G. By the theorem of Milnor [1], we may choose $K(\pi,2)$ to be an abelian topological group. Hence the homotopy classes of maps $B \longrightarrow B_G$ may be given a group structure.

(b) Given two principal G-bundles $p_1 : E_1 \longrightarrow B$ and $p_2 : E_2 \longrightarrow B$, we may form the principal $G \times G$-bundle $p_1 \times p_2 : E_1 \times E_2 \longrightarrow B \times B$; let $p' : E' \longrightarrow B$ denote the $G \times G$-bundle over B induced by the diagonal map $d : B \longrightarrow B \times B$. Consider the homomorphism $\lambda : G \times G \longrightarrow G$ defined by $\lambda(g_1, g_2) = g_1 \cdot g_2$ (recall that G is abelian !). Then by definition the λ-extension† of $p' : E' \longrightarrow B$ is the sum of the two given bundles.

LEMMA. These two methods of forming the sum of principal G-bundles are equivalent.

PROOF. Let $f_1 : B \longrightarrow B_G$ and $f_2 : B \longrightarrow B_G$ be the classifying maps for $p_1 : E_1 \longrightarrow B$ and $p_2 : E_2 \longrightarrow B$ respectively. Then the classifying map for $p_1 \times p_2 : E_1 \times E_2 \longrightarrow B \times B$ is the map

$$f_1 \times f_2 : B \times B \longrightarrow B_G \times B_G$$

since $B_{G \times G} = B_G \times B_G$, etc. The classifying map for the bundle $p' : E' \longrightarrow B$ is $(f_1 \times f_2) \circ d = (f_1, f_2) : B \longrightarrow B_G \times B_G$. The homomorphism $\lambda : G \times G \longrightarrow G$ induces the so-called "characteristic" map $\rho(\lambda) : B_{G \times G} \longrightarrow B_G$ and the classifying map of the λ-extension of $p' : E' \longrightarrow B$ is

$$\rho(\lambda) \circ (f_1, f_2) : B \longrightarrow B_G \quad .$$

However, it is seen without difficulty that $\rho(\lambda) : B_G \times B_G \longrightarrow B_G$ is homotopic to multiplication defined in B_G by the group structure (which is unique up to a homotopy). Thus $\rho(\lambda) \circ (f_1, f_2)$ is the homotopy class of the sum of f_1 and f_2 according to definition (a), q.e.d.

If $\eta_i = (E_i, B, G, p_i)$, $i = 1, 2$, are principal G-bundles over B, we will denote their sum by $\eta_1 + \eta_2$.

Recall that any cohomology class $c \in H^q(B_G)$ defines a "characteristic class" $c(\eta) \in H^q(B)$ for any principal G-bundle $\eta = (E, B, G, p)$ according to the rule

$$c(\eta) = f^*(c)$$

---

† If $p_1 : E_1 \longrightarrow B$ and $p_2 : E_2 \longrightarrow B$ are principal bundles with groups G_1 and G_2 respectively, and $\lambda : G_1 \longrightarrow G_2$ is a continuous homomorphism, then the second bundle is a λ-extension of the first if and only if there exists a map $\emptyset : E_1 \longrightarrow E_2$ such that

(a) $p_2 \emptyset = p_1$

(b) For any $x \in E_1$ and $g \in G_1$,

$$\emptyset(x \cdot g) = (\emptyset x) \cdot (\lambda g) \quad .$$

It is easily proved that given $p_1 : E_1 \longrightarrow B$ with group G_1 and the continuous homomorphism $\lambda : G_1 \longrightarrow G_2$, there exists the λ-extension $p_2 : E_2 \longrightarrow B$ which is unique up to equivalence; see Borel and Hirzebruch, [1], §6.5.

where $f : B \longrightarrow B_G$ is the classifying map for η. It follows readily from definition (a) of the sum of G-bundles that

(1) $$c(\eta_1 + \eta_2) = c(\eta_1) + c(\eta_2) \quad .$$

Next, we will consider principal G-bundles over a connected base space B such that $\pi_2(B) = 0$. Then $\pi_1(E)$ is a group extension of $\pi_1(G)$ by $\pi_1(B)$ (this follows from the homotopy exact sequence of the given bundle). Since G is connected, $\pi_1(B)$ acts trivially on $\pi_1(G)$.

THEOREM 1. If η and θ are principal G-bundles over the base space B with $\pi_2(B) = 0$, then the group extension

$$0 \longrightarrow \pi_1(G) \longrightarrow \pi_1(E_{\eta+\theta}) \longrightarrow \pi_1(B) \longrightarrow 0$$

is the Baer sum of the extensions

$$0 \longrightarrow \pi_1(G) \longrightarrow \pi_1(E_\eta) \longrightarrow \pi_1(B) \longrightarrow 0$$

and

$$0 \longrightarrow \pi_1(G) \longrightarrow \pi_1(E_\theta) \longrightarrow \pi_1(B) \longrightarrow 0 \quad .$$

PROOF. To prove this, one uses definition (b) of the sum of two G-bundles, checking that each step of the construction of the sum of the bundles corresponds to a step in the construction of the Baer sum of two group extensions. The first step gives rise to the following exact sequence

$$0 \longrightarrow \pi_1(G \times G) \longrightarrow \pi_1(E_\eta \times E_\theta) \longrightarrow \pi_1(B \times B) \longrightarrow 0$$

which may be re-written

$$0 \rightarrow \pi_1(G) \times \pi_1(G) \rightarrow \pi_1(E_\eta) \times \pi_1(E_\theta) \rightarrow \pi_1(B) \times \pi_1(B) \rightarrow 0$$

The second step gives the following diagram:

$$\begin{array}{ccccccc} & & \pi_1(E_\eta) \times \pi_1(E_\theta) & \longrightarrow & \pi_1(B) \times \pi_1(B) & \longrightarrow & 0 \\ & \nearrow & \nearrow & & \nearrow & & \\ 0 \longrightarrow \pi_1(G) \times \pi_1(G) & & & & & & \\ & \searrow & & & & & \\ & & \pi_1(E') & \longrightarrow & \pi_1(B) & \longrightarrow & 0 \end{array} \quad .$$

Here the vertical arrows denote "diagonal" maps. From the third step we get the diagram

$$\begin{array}{ccccccc} 0 \longrightarrow & \pi_1(G) \times \pi_1(G) & \longrightarrow & \pi_1(E') & \searrow & & \\ & \downarrow \lambda_* & & \downarrow & & \pi_1(B) \longrightarrow & 0 \\ 0 \longrightarrow & \pi_1(G) & \longrightarrow & \pi_1(E_{\eta+\theta}) & \nearrow & & \end{array}$$

We leave it to the reader to check that these diagrams actually correspond to the various steps in the construction of the Baer sum of two group extensions.

q.e.d.

Thus we see that we have a natural homomorphism from the group of all principal G-bundles over B into the Baer group of extensions of $\pi_1(G)$ by $\pi_1(B)$ (with $\pi_1(B)$ operating trivially on $\pi_1(G)$).

THEOREM 2. If B is an Eilenberg-MacLane space $K(\Pi, 1)$ then this natural homomorphism is an isomorphism onto.

PROOF. Let $c \in H^2(B_G,\pi)$ denote the so-called fundamental class (here $\pi = \pi_1(G) = \pi_2(B_G)$). Then it is well known that the operation of assigning to any principal G-bundle η over B the cohomology class $c(\eta) \in H^2(B,\pi)$ is an isomorphism of the group of all G-bundles over B onto $H^2(B,\pi)$. It is also well-known that $c(\eta)$ is the obstruction to a cross section of η. Using this remark, we can easily show as follows that the natural homomorphism is an isomorphism. Let $\eta = (E,B,G,p)$ be a principal G-bundle such that the extension

$$0 \longrightarrow \pi_1(G) \longrightarrow \pi_1(E) \xrightarrow{p_*} \pi_1(B) \longrightarrow 0$$

corresponds to the zero element of the Baer group, i.e., the extension splits. Let $\varphi : \pi_1(B) \longrightarrow \pi_1(E)$ be a splitting homomorphism, i.e.,

$$p_* \circ \varphi = \text{identity}.$$

Since $B = K(\Pi,1)$ and $G = K(\pi,1)$, it follows that $E = K(\pi_1(E),1)$. Therefore there exists a continuous map

$$f : B \longrightarrow E$$

such that $f_* = \varphi : \pi_1(B) \longrightarrow \pi_1(E)$. Since $B = K(\Pi,1)$, it follows readily that $p \circ f$ is homotopic to the identity map $B \longrightarrow B$. By use of the covering homotopy theorem, any homotopy of $p \circ f$ with the identity can be covered by a homotopy of f to a cross section of η. Therefore the characteristic class $c(\eta)$ vanishes, and η is a trivial bundle.

Next, we will prove that the natural homomorphism is onto. Let

$$0 \longrightarrow \pi_1(G) \xrightarrow{\psi} G \xrightarrow{\varphi} \pi_1(B) \longrightarrow 0$$

be any group extension of $\pi_1(G)$ by $\pi_1(B)$ with $\pi_1(B)$ operating trivially on $\pi_1(G)$. Let E be a space of type $K(G,1)$. As before, there exists a map $f : E \longrightarrow B$ such that $f_* = \varphi : \pi_1(E) \longrightarrow \pi_1(B)$. By use of standard techniques, we may assume that f is a fibre map in the sense of Serre (see Serre, [1], §6.1). It is readily seen that the fibre is a space of type $K(\pi_1(G),1)$. We may then replace this fibre space by a homotopically equivalent one $p' : E' \longrightarrow B$ with fibre G having all the desired properties. q.e.d.

Combining this isomorphism with the isomorphism of $H^2(B,\pi)$ with the group of G-bundles over B, we obtain a natural isomorphism between the Baer group and $H^2(\Pi,1;\pi)$. Presumably this isomorphism is the same as the usual one such as is described in Cartan and Eilenberg, [1], Chapter XIV, although I do not see how to prove this.

§3. An Application. Let B be a compact, connected 2-manifold which is not homeomorphic to the 2-sphere or real projective plane and let G be a circle group. Then B is a space of type $K(\pi_1(B),1)$ and G is a space of type $K(Z,1)$, so Theorem 2 is applicable; the group of all principal G-bundles over B is in 1-1 correspondence with the Baer group of extensions over $\pi_1(B)$ by Z, with $\pi_1(B)$ operating trivially on Z; in particular, it is isomorphic to $H^2(\pi_1(B),1,Z) = H^2(B,Z)$. The correspondence is obtained by letting the extension

$$0 \longrightarrow \pi_1(G) \longrightarrow \pi_1(E) \longrightarrow \pi_1(B) \longrightarrow 0$$

correspond to the characteristic class $c \in H^2(B,Z)$ for any principal G-bundle $p : E \longrightarrow B$. In this case, it is well known that $H^2(B,Z) = Z$ or Z_2, depending on whether B is orientable or not. Since the group $H^2(B,Z)$ admits at most one non-trivial automorphism, we see that the isomorphism

$$\text{Baer Group} \longleftrightarrow H^2(B,Z) = H^2(\pi_1(B),1,Z)$$

which we have just described and the usual isomorphism which is defined between these two groups (e.g. Eilenberg-Cartan, [1], Chapter XIV) are the same or the negatives of each other. This fact may be used to give a purely algebraic description of the fundamental group $\pi_1(E)$.

An interesting special case occurs when B is an orientable surface of Euler characteristic $\chi \leq 0$ and $p : E \longrightarrow B$ denotes the bundle of unit tangent vectors. In this case we have the following theorem describing the structure of $\pi_1(E)$:

THEOREM 3. Under the hypotheses just mentioned the group extension

$$0 \longrightarrow Z \longrightarrow \pi_1(E) \xrightarrow{p^*} \pi_1(B) \longrightarrow 0$$

is that which is determined by one of the elements $\pm \chi \cdot u \in H^2(\pi_1(B),Z)$, where u denotes a generator of the infinite cyclic group $H^2(\pi_1(B),Z)$.

CHAPTER IV.

MINIMAL FLOWS ON NILMANIFOLDS

by

L. Auslander
F. Hahn
L. Markus

§1. A Classical Example. In this example let G be the real n-dimensional vector space R^n, $n \geq 1$, and let H be the discrete subgroup of all points having integral coordinates. Then $M = G/H$ is the n-torus. Choose a Euclidean metric in G and this projects to a locally flat Riemann metric on M.

Consider a one-parameter subgroup

$$\varphi\colon\ T \longrightarrow G\colon\ t \longrightarrow \varphi(t)$$

which is specified by an initial tangent vector v at the origin of G, and let

$$\varphi^*\colon\ M \times T \longrightarrow M\colon\ (x,t) \longrightarrow \varphi^*_t(x)$$

be the corresponding induced flow on the torus M. Each transformation φ^*_t is an isometry of M onto M, and the flow φ^* has many interesting geometric properties.

a) The one-parameter groups $\varphi(t)$ and $\varphi(\lambda t)$, for a real $\lambda \neq 0$, corresponding to the initial vectors v and λv, define topologically isomorphic transformation groups on M. However, there are only a countable number (Fomin [1]) of unit vectors at the origin of G which induce a transformation group on M which is topologically isomorphic to that induced by v. Thus there exist a continuum of topologically different transformation groups on the torus M which are induced by one-parameter subgroups of $G = R^n$.

b) The flow φ^* is distal, that is: for each pair of distinct points x and y of M the distance

$$d(\varphi^*_t(x), \varphi^*_t(y)) > \varepsilon > 0$$

for some $\varepsilon > 0$ and for all $t \in T$.

c) The flow φ^* is pointwise almost periodic (Nemytskii-Stepanov [1]), that is: for each point c and for each neighborhood N of x there exists a relatively dense real sequence $\{t_k\}_{k=-\infty}^{\infty}$ such that $\varphi^*_{t_k}(x) \in N$ for $k = 0, \pm 1, \pm 2, \ldots$.

d) The flow φ^* is equicontinuous on M, that is: for each point $x \in M$ and for each $\varepsilon > 0$ there exists a $\delta > 0$ such that

$$d(y,x) < \delta$$

implies

$$d(\varphi_t^*(y), \varphi_t^*(x)) < \varepsilon$$

for all $t \in T$.

e) There exists a unit vector v at the origin of G for which the induced transformation group

$$\varphi^*: M \times T \longrightarrow M$$

is minimal (Koksma [1]), that is: for each point $x \in M$ the orbit

$$\{\varphi_t^*(x) \mid \text{ for } t \in T\}$$

is dense in M. In fact, the set of unit vectors which define such a minimal flow includes all but a set of the first category in the sphere of unit vectors. Thus there exist a continuum of topologically different minimal flows on the torus which are induced by one-parameter subgroups of G.

§2. Principal Results. The object of this paper is to extend the study of flows induced by one-parameter subgroups from tori, which are compact homogeneous spaces of abelian Lie groups, to nilmanifolds, which are compact homogeneous spaces of nilpotent Lie groups. Our main result is that all the properties a), b), c), and e) hold for nilmanifolds. It is shown that such a minimal flow on a nilmanifold, other than a torus, cannot be d) equicontinuous at even one point. Thus we provide an infinite number of minimal flows, under the real line T, which are distal yet not equicontinuous, in answer to a question of R. Ellis [1].

§3. Uniform Discrete Subgroups and Canonical Coordinates. In this section we describe the basic results on nilmanifolds and uniform discrete subgroups of nilpotent Lie groups, as proved by A. Malcev [1].

DEFINITION. Let $x_1(t), x_2(t), \ldots, x_n(t)$ be one-parameter subgroups of a connected, simply-connected nilpotent Lie group G. Require that

1) the map $(t_1, t_2, \ldots, t_n) \longrightarrow g = x_1(t_1) \cdot x_2(t_2) \ldots x_n(t_n)$ is a diffeomorphism of R^n onto G.
2) each subset $\{x_i(t_i) \cdot x_{i+1}(t_{i+1}) \ldots x_n(t_n)\}$ for $i = 1, \ldots, n$ is a closed normal subgroup G_i of G.
3) the factor groups G_i/G_{i+1}, $i = 1, \ldots, n$ (and $G_{n+1} = e$) are each R^1.

Then $\{x_1(t), \ldots, x_n(t)\}$ define a system of canonical coordinates of the second kind on G.

Now consider canonical coordinates of the second kind in G.

If $u_i(t)$, $i = 1,\ldots,n$ are the canonical coordinates of a one-parameter subgroup on G, then each $u_i(t)$ is a polynomial in t. Again write canonical coordinates of the second kind for points $x = (\xi_1,\ldots,\xi_n)$ and $y = (\eta_1,\ldots,\eta_n)$ with product $z = xy = (\zeta_1,\ldots,\zeta_n)$. Then

$$\zeta_i = \xi_i + \eta_i + Q_i(\xi_1,\ldots,\xi_{i-1},\eta_1,\ldots,\eta_{i-1})$$

for $i = 1,\ldots,n$ where Q_i are real polynomials.

REMARKS. It is easy to see that $Q_1 \equiv 0$ and also $Q_i(\xi_1,\ldots,\xi_{i-1},0,0,\ldots,0) \equiv 0$ for $i = 1,\ldots,n$. For instance, let

$$x_1(\zeta_1)\ldots x_n(\zeta_n) = x_1(\xi_1)\ldots x_i(\xi_i)x_{i+1}(\xi_{i+1})\ldots x_n(\xi_n)x_i(\eta_i)\ldots x_n(\eta_n)$$

or

$$x_1(\zeta_1)\ldots x_i(\zeta_i)g_{i+1} = x_1(\xi_1)\ldots x_i(\xi_i + \eta_i)g'_{i+1} ,$$

where g_{i+1} and g'_{i+1} belong to G_{i+1}. Now write the unique expansion

$$g'_{i+1}\, g_{i+1}^{-1} = x_{i+1}(\tau_{i+1})\ldots x_n(\tau_n) .$$

Thus by uniqueness,

$$\zeta_i = \xi_i + \eta_i$$

and hence $Q_i(\xi_1,\ldots,\xi_{i-1},0,\ldots,0) \equiv 0$ for each $i = 1,\ldots,n$.

DEFINITION. Let D be a discrete group. A canonical basis for D is a finite collection of elements $\{d_1,\ldots,d_n\}$ of D such that:

1) each $d \in D$ can be represented uniquely by $d = d_1^{m_1} d_2^{m_2}\ldots d_n^{m_n}$ for integers $m_1,m_2,\ldots,m_n$.
2) each collection $\{d_i^{m_i}\ldots d_n^{m_n}\}$ forms an invariant subgroup D_i of D, for $i = 1,\ldots,n$.
3) each quotient group D_i/D_{i+1} is Z, for $i = 1,\ldots,n$ and $D_{n+1} = e$.

In his penetrating work on nilmanifolds A. Malcev [1] proves the following basic results.

LEMMA. Let D be a uniform discrete subgroup of a connected, simply-connected, nilpotent Lie group G. Let the lower central series of G be

$$G = G^0 \supset G^1 \supset \ldots \supset G^\ell = e .$$

Then $D_i = G^i \cap D$ is a uniform discrete subgroup of G^i and the image of D in G/G^i, $i = 1,\ldots,\ell$, is a uniform discrete subgroup of G/G^i. Let $d_1,\ldots,d_s$ in D project to a canonical basis for D/D_ℓ and let $d_{s+1},\ldots,d_n$ be any basis for the abelian group D_ℓ. Then $\{d_1,\ldots,d_s,d_{s+1},\ldots,d_n\}$ is a canonical basis for D. Thus D always has a canonical basis.

LEMMA. Let D be a uniform discrete subgroup of a connected, simply-connected, nilpotent Lie group G. Let $\{d_1,\ldots,d_n\}$ be a canonical basis for D. Then there exists a system of canonical coordinates of the second kind for G, $\{x_1(t),\dots,x_n(t)\}$ such that $x_i(1) = d_i$, for $i = 1,\dots,n$.

THEOREM. Let $M = G/D$ and $\hat{M} = \hat{G}/\hat{D}$ where D and $\hat{D}$ are uniform discrete subgroups of the connected, simply-connected, nilpotent Lie groups G and $\hat{G}$, respectively. Let

$$\Phi : D \longrightarrow \hat{D}$$

be an isomorphism of D onto $\hat{D}$. Then there exists a unique diffeomorphism-isomorphism of G onto $\hat{G}$, which restricts to Φ on D.

COROLLARY. Nilmanifolds with isomorphic fundamental groups are diffeomorphic.

§4. An Example. The unique nilmanifold, up to diffeomorphism, of dimension 1 is the circle S^1. The unique nilmanifold of dimension 2 is the torus $S^1 \times S^1$.

In dimension 3 there are two nilpotent real Lie algebras: the commutative Lie algebra, which yields only the torus $S^1 \times S^1 \times S^1$ as a nilmanifold, and the Lie algebra $\mathfrak{N}(3,R)$ of class 2 generated by the basis $\{e_1, e_2, e_3\}$ satisfying

$$[e_1,e_2] = e_3, \quad [e_1,e_3] = 0, \quad [e_2,e_3] = 0 \ .$$

Representing

$$e_1 = \begin{pmatrix} 0 & 1 & 0 \\ 0 & 0 & 0 \\ 0 & 0 & 0 \end{pmatrix}, \quad e_2 = \begin{pmatrix} 0 & 0 & 0 \\ 0 & 0 & 1 \\ 0 & 0 & 0 \end{pmatrix}, \quad e_3 = \begin{pmatrix} 0 & 0 & 1 \\ 0 & 0 & 0 \\ 0 & 0 & 0 \end{pmatrix},$$

we use the exponential map and compute the corresponding connected, simply-connected, nilpotent Lie group to be T_3, the set of all real matrices of the form

$$\begin{pmatrix} 1 & x & z \\ 0 & 1 & y \\ 0 & 0 & 1 \end{pmatrix}$$

The vectors $\{e_1, e_2, e_3\}$ define a canonical system of coordinates of the first kind, and the one-parameter subgroups

$$x_1(t) = \begin{pmatrix} 1 & t & 0 \\ 0 & 1 & 0 \\ 0 & 0 & 1 \end{pmatrix}, \quad x_2(t) = \begin{pmatrix} 1 & 0 & 0 \\ 0 & 1 & t \\ 0 & 0 & 1 \end{pmatrix}, \quad x_3(t) = \begin{pmatrix} 1 & 0 & t \\ 0 & 1 & 0 \\ 0 & 0 & 1 \end{pmatrix}$$

yield the corresponding system of canonical coordinates of the second kind.

The lower central series of T_3 is

$$T_3 = G^0 \supset G^1 \supset e$$

where the commutator subgroup G^1 is exactly the center and consists of the one-parameter subgroup $x_3(t)$.

Let D be a uniform discrete subgroup of T_3 having a canonical basis d_1, d_2, d_3. Choose this basis so that d_1 and d_2 project to a canonical basis for the image of D in G/G^1, and d_3 is a basis for the commutative group $G^1 \cap D$. Thus

$$d_1 = \begin{pmatrix} 1 & x_1 & z_1 \\ 0 & 1 & y_1 \\ 0 & 0 & 1 \end{pmatrix}, \quad d_2 = \begin{pmatrix} 1 & x_2 & z_2 \\ 0 & 1 & y_2 \\ 0 & 0 & 1 \end{pmatrix}, \quad d_3 = \begin{pmatrix} 1 & 0 & z_3 \\ 0 & 1 & 0 \\ 0 & 0 & 1 \end{pmatrix}$$

and $x_1 y_2 \neq x_2 y_1$, $z_3 \neq 0$.

Then

$$d_1 d_2 d_1^{-1} d_2^{-1} = d_3^k$$

for some integer $k \neq 0$. Choose d_3 so that $k \geq 1$. Now the commutation relations

$$\begin{aligned} [d_1, d_2] &= d_1 d_2 d_1^{-1} d_2^{-1} = d_3^k \\ [d_1^{-1}, d_2^{-1}] &= d_3^k \\ [d_1, d_2^{-1}] &= d_3^{-k} \\ [d_1^{-1}, d_2] &= d_3^{-k} \end{aligned}$$

and the fact that d_3 is in the center of D, determine the group D up to isomorphism; since there are no relations in D other than those generated by the above commutation relations.

But for each $k = 1,2,3,\ldots$ the corresponding discrete group $D(k)$ is different since the center of $D(k)$ modulo the commutator subgroup of $D(k)$ is exactly $\mathbf{Z}_k$. Thus each uniform discrete subgroup of T_3 is isomorphic to exactly one $D(k)$, $k = 1,2,3,\ldots$ which is generated by

$$\delta_1 = \begin{pmatrix} 1 & 1 & 0 \\ 0 & 1 & 0 \\ 0 & 0 & 1 \end{pmatrix}, \quad \delta_2 = \begin{pmatrix} 1 & 0 & 0 \\ 0 & 1 & 1 \\ 0 & 0 & 1 \end{pmatrix}, \quad \delta_3 = \begin{pmatrix} 1 & 0 & \frac{1}{k} \\ 0 & 1 & 0 \\ 0 & 0 & 1 \end{pmatrix}.$$

Using the commutation relations obtained above, it is easy to see that each element of $D(k)$ can be written uniquely as

$$\delta_1^a \, \delta_2^b \, \delta_3^c = \begin{pmatrix} 1 & a & ab + \frac{c}{k} \\ 0 & 1 & b \\ 0 & 0 & 1 \end{pmatrix} \quad \text{for integers } a,b,c.$$

Then a straight forward computation shows that $D(k)$ is a uniform discrete subgroup of T_3.

Therefore each 3-dimensional nilmanifold M^3 is diffeomorphic with just one $T_3/D(k)$, which has a nonabelian fundamental group $D(k)$, or else M^3 is the torus $S^1 \times S^1 \times S^1$.

Consider the uniform discrete subgroups $D(k)$ of T_3 and form the direct product of $D(k)$ with $n \geq 0$ integer groups $D(k,n) = D(k) \times \mathbf{Z} \times \ldots \times \mathbf{Z}$. Then $D(k,n)$ is a uniform discrete subgroup of $T_3 \times R^n$ and $M(k,n) = T_3 \times R^n/D(k,n)$ is a nilmanifold of dimension $3 + n$ having

$D(k,n)$ as fundamental group. The center modulo the commutator subgroup of $D(k,n)$ has the torsion coefficient of k. Hence $D(k,n)$ is not isomorphic with $D(k',n)$ for $k \neq k'$. Hence the infinitely many nilmanifolds $M(k,n)$, for a fixed $n \geq 0$ and $k = 1,2,3,\ldots,$ are of different homotopy types.

§5. Winding Class as a Topological Invariant. Let $M = G/D$ be a nilmanifold where G is a connected, simply-connected, nilpotent Lie group and D is a uniform discrete subgroup of G. Let $T = R^1$ be the additive group of real numbers and let

$$\varphi : T \longrightarrow G : t \longrightarrow \varphi(t) \in G$$

be a one-parameter subgroup in G. Then

$$\varphi^*: M \times T \longrightarrow M : (gD,t) \longrightarrow \varphi_t^*(gD) = (\varphi(t)g)D$$

defines a flow on M, the left coset space of G/D. We say that φ^* is the flow on M induced by φ (or by the initial tangent vector $v = \dot{\varphi}(0)$ of $\varphi(t)$) on G. Different one-parameter subgroups of G induce different flows on M.

The collection F of flows on M, induced by one-parameter subgroups of G, is a topological invariant of M. That is, choose a different homeomorphism of M with a left coset space $\hat{G}/\hat{D}$, where $\hat{G}$ is necessarily isomorphic with G so that $\hat{D}$ is mapped onto D. The resulting collection $\hat{F}$ of flows on M, induced by one-parameter subgroups of $\hat{G}$, is topologically equivalent to F by a homeomorphism of M onto itself.

For the remainder of this section we fix a diffeomorphism of M with G/D. To each induced flow φ^* on M, arising from φ on G, we shall assign a classical flow φ^τ on an s-torus, $s \geq 2$. In the case where M is a torus, we shall have $\varphi^* = \varphi^\tau$. Let $M = G/D$ be a nilmanifold of dimension $n \geq 2$, as above. Let G^1 be the commutator subgroup of G and define $D_1 = D \cap G^1$. Then D_1 is normal in D and, by the natural projection $G \longrightarrow G/G^1$, there is defined an isomorphism of D/D_1 onto a uniform discrete subgroup D_* of G/G^1. Also $(G/G^1)/D_*$ is a torus group τ^s of dimension $s \geq 2$.

DEFINITION. Let $M = G/D$ be a nilmanifold of dimension $n \geq 2$, as above. Let the one-parameter subgroup φ on G induce the flow φ^* on M. Consider the torus $\tau^s = (G/G^1)/D_*$, as above. Define the flow φ^τ on τ^s by

$$\varphi^\tau: \tau^s \times T \longrightarrow \tau^s: (hD_*,t) \longrightarrow (\varphi'(t)h)D_* ,$$

where $h = gG^1 \in G/G^1$ and $\varphi'(t)h = (\varphi(t)g)G^1$.

It will be shown in section 7 that each induced flow φ^* on a nilmanifold $M = G/D$ is pointwise almost periodic, and thus recurrent as $t \longrightarrow +\infty$ and as $t \longrightarrow -\infty$ at each point $Q \in M$. Using this knowledge, we now define the winding class of φ^*. The winding class extends the classical concept of the winding number (or rotation number) for a flow on a torus surface, cf. Poincaré [1], Schwartzman [1].

REMARKS. Let $M = G/D$ be a nilmanifold of dimension $n \geq 2$, as above. At each point $Q \in M$ we construct the fundamental group $\pi_1(M,Q)$ which is isomorphic to D. Furthermore, this isomorphism is fixed, up to an inner automorphism of $\pi_1(M,Q)$. The subgroup $D_1 = D \cap G^1$ is characteristic in D and so we call $D_1(Q)$ the corresponding subgroup of $\pi_1(M,Q)$.

Now $\pi_1(M,Q)/D_1(Q)$ is isomorphic to D/D_1, and in a canonical way, since D/D_1 is commutative. Now $\pi_1(M,Q)/D_1(Q)$ is a $\mathbf{Z}$-module and we form $R^1 \otimes_{\mathbf{Z}} \pi_1(M,Q)/D_1(Q)$ which we further consider as an R^1-module. Since $R^1 \otimes \pi_1(M,Q)/D_1(Q)$ is a real vector space of dimension $s \geq 2$, we define the space P_Q^{s-1} of all one-dimensional subspaces of $R^1 \otimes \pi_1(M,Q)/D_1(Q)$. Note that P_Q^{s-1} is diffeomorphic with the real projective space P^{s-1} of dimension $s-1 \geq 1$, once a basis is chosen in D/D_1.

DEFINITION. Let the one-parameter group $\varphi(t)$ on G induce a flow φ^* on a nilmanifold $M = G/D$ of dimension $n \geq 2$, as above. At a point $Q \in M$ choose a ball coordinate neighborhood N of Q and consider the orbit $\varphi_t^*(Q)$ on $0 \leq t < \infty$. Choose recurrence times $0 < t_1 < t_2 < \ldots < t_k < \ldots \longrightarrow \infty$ at which $\varphi_{t_k}^*(Q) \in N$. Construct the loop ℓ_k defined by the curve $\varphi_t^*(Q)$ on $0 \leq t \leq t_k$ and any arc in N joining $\varphi_{t_k}^*(Q)$ to Q. Then ℓ_k determines an element of $\pi_1(M,Q)$ and hence an element $\bar{\ell}_k$ in $\pi_1(M,Q)/D_1(Q) \cong D/D_1$. If $\bar{\ell}_k \neq 0$, it determines a point $\alpha_k(Q)$ in P_Q^{s-1}, the line space of $R^1 \otimes D/D_1$.

If $\bar{\ell}_k$ is never zero for large k, and if $\lim_{k \to \infty} \alpha_k(Q) = \alpha(Q)$ exists in P_Q^{s-1}, then $\alpha(Q)$ is independent of the choice of neighborhood N and the recurrence times $\{t_k\}$, and $\alpha(Q)$ is called the winding class of φ^* at Q as $t \longrightarrow +\infty$.

If $\bar{\ell}_k = 0$ for all large k, then we say that the winding class of φ^* at Q as $t \longrightarrow +\infty$ is zero. Similar definitions hold for $t \longrightarrow -\infty$.

REMARKS. If a basis is chosen such that $D/D_1 \cong \mathbf{Z}^s$, and if $\alpha(Q)$ exists as $t \longrightarrow +\infty$, then $\alpha(Q)$ determines a point in the real projective space P^{s-1} and hence we can assign s real homogeneous coordinates to $\alpha(Q)$. Two different bases in D/D_1 determine homogeneous coordinates for $\alpha(Q)$ which are transforms of one another by an integral matrix with determinant ± 1. For a flow with zero winding class we assign the homogeneous coordinates $(0,0,\ldots,0)$, for every basis of D/D_1.

For a classical flow φ^τ on a torus τ^s, $s \geq 2$, the winding class $\alpha(Q)$ exists at each point $Q \in \tau^s$ as $t \longrightarrow +\infty$ and as $t \longrightarrow -\infty$. Also $\alpha(Q)$ is zero if and only if the flow is the zero flow. A choice of basis for $\pi_1(\tau^s,Q)$ defines a basis for $\pi_1(\tau^s,Q_1)$ at every point $Q_1 \in \tau^s$.

It is easily seen that, if $\alpha(Q) \neq 0$, the homogeneous coordinates of $\alpha(Q_1)$ are the same for all $Q_1 \in \tau^S$ as $t \longrightarrow +\infty$ and $t \longrightarrow -\infty$ (and they are all the same for the two time senses), once the basis for $\pi_1(\tau^S)$ is designated. For $s = 2$, the winding class $\alpha(Q) = \alpha$ in homogeneous coordinates, is the Poincaré [1] winding number of the flow. Two classical flows φ_1^τ and φ_2^τ on τ^S are known (Fomin [1]) to be topologically equivalent if and only if their homogeneous coordinates are related by an integral matrix with determinant ± 1.

THEOREM 1. Let $\varphi(t)$ be a one-parameter subgroup on a connected, simply-connected, nilpotent Lie group G, which induces a flow φ^* on a nilmanifold $M = G/D$ of dimension $n \geq 2$, as above. Then the winding class $\alpha(Q)$ of φ^* exists at each point $Q \in M$, as $t \longrightarrow +\infty$ and as $t \longrightarrow -\infty$. Given a basis for D/D_1, a corresponding basis is determined in each $\pi_1(M,Q)/D_1(Q)$ and also in $\pi_1(\tau^S)$, where $\tau^S = (G/G^1)/D_*$, as above. If $\alpha(Q) = 0$ for $t \longrightarrow +\infty$, then $\alpha(Q_1) = 0$ as $t \longrightarrow +\infty$ and as $t \longrightarrow -\infty$ for every point $Q_1 \in M$. If $\alpha(Q) \neq 0$ as $t \longrightarrow +\infty$, then the homogeneous coordinates of $\alpha(Q_1)$ are the same for $t \longrightarrow +\infty$ and for $t \longrightarrow -\infty$, and $\alpha(Q_1) = \alpha(Q)$ for all $Q_1 \in M$, when referred to the corresponding bases. Moreover, the homogeneous coordinates of $\alpha(Q)$ are those of the torus flow φ^τ on τ^S, referred to the corresponding basis.

PROOF. Given a basis in $D/D_1 \cong \mathbf{Z}^S$, we determine a basis in each $\pi_1(M,Q)/D_1(Q)$ for $Q \in M$ and in $\pi_1(\tau^S)$, as discussed above. Considering G as the universal covering space of M, we note that each right multiplication of G by an element of D determines an element of D/D_1 and thus s homogeneous coordinates. The distinguished basis for $\pi_1(\tau^S)$ is defined as follows.

Each element $d \in D$ defines a right translation of G, which belongs to $\pi_1(M)$, and also the coset dG^1 defines a translation of G/G^1 which belongs to $\pi_1(\tau^S)$. Two elements of D, which are in the same coset of D/D_1, yield the same element of $\pi_1(\tau^S)$. In this way $\pi_1(\tau^S)$ is isomorphic with $\mathbf{Z}^S$.

Consider a closed loop ℓ in M, based at $Q \in M$. Let $\tilde{Q}_1$ and $\tilde{Q}_2$ in G lie above Q that is, the cosets $\tilde{Q}_1 D = \tilde{Q}_2 D = Q$, and say that ℓ lifts to $\tilde{\gamma}$ joining $\tilde{Q}_1$ to $\tilde{Q}_2$ in G. Now $\tilde{Q}_1 = \tilde{Q}_2 d_{12}$ with $d_{12} \in D$ and so $\tilde{Q}_1 G^1 = \tilde{Q}_2 d_{12} G^1$ or $\tilde{Q}_1 G^1 = (\tilde{Q}_2 G^1) \cdot (d_{12} G^1)$. Thus $\tilde{Q}_1$ and $\tilde{Q}_2$ project to the same point Q^τ in $\tau^S = (G/G^1)/D_*$. Thus the curve $\tilde{\gamma}$ projects to a closed loop ℓ^τ in τ^S, based at Q^τ. Moreover the loop ℓ^τ in τ^S has precisely the same homogeneous coordinates as has ℓ; and ℓ^τ is null homotopic in τ^S just in case ℓ determines an element of $D_1(Q)$.

Return to the consideration of the one-parameter group $\varphi(t)$ in G and the induced flows φ^* in M and φ^τ in τ^S. The recurrence times for φ^* near Q are a subsequence of the recurrence times of φ^τ near Q^τ in τ^S. Thus our construction shows that the winding class of φ^* at $Q \in M$ exists in P_Q^{S-1} as $t \longrightarrow +\infty$ and has the same homogeneous coordinates as has φ^τ at $Q^\tau \in \tau^s$; or else φ^* has zero winding class and φ^τ is the zero velocity flow on τ^s.

But the winding class of φ^τ is independent of the point Q^τ and is the same for $t \longrightarrow +\infty$ and for $t \longrightarrow -\infty$. Thus the winding class for φ^* at each point $Q \in M$ for $t \longrightarrow \pm\infty$ has the same homogeneous coordinates as has φ^τ on τ^s. q.e.d.

LEMMA. Let $\varphi_1(t)$ and $\varphi_2(t)$ be one-parameter subgroups of G, and let φ_1^* and φ_2^* be the corresponding induced flows on the nilmanifold $M = G/D$ of dimension $n \geq 2$, and let φ_1^τ and φ_2^τ be the corresponding induced flows on τ^s, $s \geq 2$, as above. If

$$\varphi_1^\tau : \tau^s \times T \longrightarrow \tau^s$$

and

$$\varphi_2^\tau : \tau^s \times T \longrightarrow \tau^s$$

are topologically inequivalent flows, then

$$\varphi_1^* : M \times T \longrightarrow M$$

and

$$\varphi_2^* : M \times T \longrightarrow M$$

are topologically inequivalent flows.

PROOF. Again choose a basis in D/D_1 and hence corresponding bases in $\pi_1(M,Q)/D_1(Q)$ at each point $Q \in M$ and in $\pi_1(\tau^s)$, as above. Since φ_1^τ and φ_2^τ are topologically inequivalent, they are not both zero, and the homogeneous coordinates for φ_1^τ are not obtained from those of φ_2^τ by an integral matrix transformation with determinant ± 1.

Therefore the homogeneous coordinates of φ_1^* and of φ_2^* are not related by an integral matrix transformation with determinant ± 1.

If there did exist a homeomorphism ψ of M onto M (and a change of time scale if necessary) which carries φ_1^* to φ_2^*, then ψ would carry $\pi_1(M,Q)/D_1(Q)$ to $\pi_1(M,\psi(Q))/D_1(\psi(Q))$. Thus ψ would define a change of basis in D/D_1. But this would imply that the homogeneous coordinates of φ_1^* are the same as those of φ_2^*, when referred to the two given bases in D/D_1. In this case, when referred to the same basis in D/D_1, the homogeneous coordinates of φ_1^* are related to those of φ_2^* by an integral matrix transformation with determinant ± 1. This is a contradiction and thus φ_1^* and φ_2^* are topologically inequivalent flows on M.

q.e.d

We shall next show that the set V of vectors at the origin of G, which induce flows on $M = G/D$ topologically equivalent to a given flow, is negligible in a certain technical sense. In fact V will be shown to have measure zero and category I in the tangent space L at the identity of G.

The sets of Lebesgue measure zero are intrinsically determined by the affine structure of L. That is, whatever constant non-singular n-form is prescribed in L, the class of sets of measure zero is the same.

Also it is interesting to note that a set V_1 of unit vectors in L, for some choice of a Euclidean metric in L, has measure zero in the unit sphere of L if and only if the saturation V of V_1, by lines through the origin of L, has measure zero in L. Also V_1 is of first category in the unit sphere of L if and only if the saturation V is of first category in L.

Since two non-zero vectors of L which lie in the same one-dimensional subspace of L yield topologically equivalent flows on M, it will often be convenient to consider only unit vectors in L.

THEOREM 2. Let $\varphi(t)$, with initial tangent vector v, be a one-parameter subgroup of a connected, simply-connected, nilpotent Lie group G. Let D be a uniform discrete subgroup of G and let φ^* be the induced flow on the nilmanifold $M = G/D$, of dimension $n \geq 2$. Let V be the set of vectors in the tangent space L at the identity of G, which determine flows on M topologically equivalent to φ^*. Then V has measure zero and category I in L. Thus there exist a non-countable infinity of topologically inequivalent flows induced on M.

PROOF. Let V be the set of vectors in L which induce flows on M topologically equivalent to φ^*. Now L is a vector space direct sum $L = L_1 \oplus L_2$, where L_1 is the Lie algebra of G^1 and L_2 is any complementary subspace. A vector $u \in L$ has a unique decomposition $u = u_1 + u_2$ with $u_1 \in L_1$ and $u_2 \in L_2$.

The component u_2 determines the torus flow $\varphi^\tau[u]$ on τ^s corresponding to the one-parameter group $\varphi(t,u)$ in G, generated by u. Thus $\varphi^\tau[u] = \varphi^\tau[u_2]$. The set of vectors $u_2 \in L_2$ for which $\varphi^\tau[u_2]$ is topologically equivalent to $\varphi^\tau[v]$ on τ^s comprise a countable collection C of one-dimensional subspaces of L_2 (or is just the origin if $v \in L_1$ so $\varphi^\tau[v]$ is the zero flow on τ^s).

Therefore the set V of all vectors in L whose projections in L_2 lie in C is a set of measure zero and of first category in L.

q.e.d.

EXAMPLES. Consider the group $G = T_3$ of all real matrices of the form

$$\begin{pmatrix} 1 & x & z \\ 0 & 1 & y \\ 0 & 0 & 1 \end{pmatrix}$$

and let D be the uniform discrete subgroup of matrices

$$\begin{pmatrix} 1 & a & c \\ 0 & 1 & b \\ 0 & 0 & 1 \end{pmatrix}$$

for all integers a,b,c. Then $M = G/D$ is a nilmanifold of dimension 3, as we have noted above. It is easy to see that a fundamental domain F for M in G is the set of matrices with $0 \le x,y,z < 1$, that is each point of M is represented by exactly one matrix of F.

Consider a one-parameter subgroup $\varphi(t)$ of G given by

$$\exp t \begin{pmatrix} 0 & \alpha & \gamma \\ 0 & 0 & \beta \\ 0 & 0 & 0 \end{pmatrix} = \begin{pmatrix} 1 & \alpha t & \gamma t + \frac{\alpha\beta}{2} t^2 \\ 0 & 1 & \beta t \\ 0 & 0 & 1 \end{pmatrix} .$$

Take a point $Q \in M$ given by the coset

$$\begin{pmatrix} 1 & x_0 & z_0 \\ 0 & 1 & y_0 \\ 0 & 0 & 1 \end{pmatrix} D .$$

The orbit $\varphi_t^*(Q)$ in M is

$$\begin{pmatrix} 1 & \alpha t + x_0 & \gamma t + \frac{\alpha\beta}{2} t^2 + z_0 + \alpha t y_0 \\ 0 & 1 & \beta t + y_0 \\ 0 & 0 & 1 \end{pmatrix} D .$$

To show that $\varphi_t^*(Q)$ is dense in M we must show that

$$f_1(\ell,m,n,t) = \alpha t + x_0 + \ell$$

$$f_2(\ell,m,n,t) = \beta t + y_0 + m$$

$$f_3(\ell,m,n,t) = \gamma t + \frac{\alpha\beta}{2} t^2 + z_0 + \alpha t y_0 + n + \alpha t m + x_0 m$$

are dense in the fundamental domain F for integers (ℓ,m,n) and real t.

We have not been able to give a constructive proof of this fact for any choice of (α,β,γ). However our theory, below, shows that the flow is minimal for almost all (α,β,γ).

Incidentally, it is easy to see that each induced φ^* is distal; since the torus flow φ^τ is an isometry and since the coordinate z_0 enters f_3 linearly. However each minimal flow φ^* is not equicontinuous, since M is not the group-space of an abelian topological group, cf. Gottshalk-Hedlund [1] p. 39. Thus these examples resolve a problem posed by R. Ellis [1]; to construct a minimal, distal flow which is not equicontinuous.

§6. The Fundamental Domain of a Nilmanifold. Let G be a connected, simply connected, nilpotent Lie group and let D be a uniform discrete subgroup so the left coset space $M = G/D$ is a compact nilmanifold. Choose a canonical basis for D and corresponding canonical coordinates of the second kind for G. Thus we define a diffeomorphism of R^n onto G

$$C : R^n \longrightarrow G : (t_1 \ldots t_n) \longrightarrow g = x_1(t_1) \cdot x_2(t_2) \ldots x_n(t_n)$$

which carries the integral lattice $\mathbf{Z}$ onto D. We shall often use the coordinates of R^n to designate the points of G.

Let $\mathfrak{G}$ be the Lie algebra of G and let $X_1, X_2, \ldots, X_n$, be a basis for $\mathfrak{G}$ such that $\exp X_i t = x_i(t)$, $i = 1,2,\ldots,n$. The mapping $\exp : \mathfrak{G} \longrightarrow G$ is a diffeomorphism onto G. We let $\log : G \longrightarrow \mathfrak{G}$ be the inverse map.

LEMMA. Let $M = G/D$ and let $\Sigma a_i X_i$ be a point in $\mathfrak{G}$. Let

$$F = \left\{ \sum_{i=1}^{n} b_i X_i \ : \ a_i \leq b_i < a_i + 1, \quad i = 1,2\ldots,n \right\} .$$

Under these conditions $\exp(F)$ is a fundamental domain for M in the sense that:

a) each point of M is represented by exactly one point of $\exp(F)$,

b) the natural projection $G \longrightarrow G/D = M$ is continuous on $\exp(F)$ and is a homeomorphism when restricted to the interior of $\exp(F)$.

PROOF. Since the natural projection $G \longrightarrow G/D = M$ is a continuous open map, we must only prove that each point of M is represented by exactly one point of $\exp(F)$. If the dimension of G is one then this statement is clearly true. Suppose that the statement is true if the dimension of G is $n-1$.

Let $\mathfrak{H}$ be the ideal in $\mathfrak{G}$ generated by X_n. Since X_n is in the center we see that $\mathfrak{H} = \{X_n t : -\infty < t < +\infty\}$. We let $H = \exp(\mathfrak{H})$ and let $G^* = G/H$ and $\mathfrak{G}^* = \mathfrak{G}/\mathfrak{H}$. We shall use asterisks to designate the image of elements of G and $\mathfrak{G}$ under the natural projection in G^* and $\mathfrak{G}^*$, respectively. We see that G^* is a connected, simply-connected, nilpotent, Lie group of dimension $n-1$ and that $G^*/D^* = M^*$ is a nilmanifold. The elements $x_1^*(1)$, $x_2^*(1) \ldots$, $x_{n-1}^*(1)$ form a canonical basis for D^* and each element $g^* \in G^*$ may be written

$$g^* = x_1^*(t_1)\, x_2^*(t_2) \ldots x_{n-1}^*(t_{n-1}) .$$

It follows from the inductive hypothesis that $\exp(F^*)$ is a fundamental domain for M^*. If $g = (\xi_1,\ldots,\xi_n)$ is any element of G then there is a unique $d_0^* \in D^*$ for which $g^* d_0^* \in \exp F^*$. Let $d_0^* = x_1^*(m_1) \ldots x_{n-1}^*(m_{n-1})$ and choose $b_0^* \in \mathfrak{G}^*$ such that $b_0^* = \sum_{i=1}^{n-1} b_i X_i^*$ and $g^* d_0^* = \exp b_0^*$. There is a unique integer m_n for which $a_n \leq \xi_n + m_n < a_n + 1$. Let $b_n = \xi_n + m_n$ and let $d = (m_1, m_2, \ldots, m_n) \in D$. Since X_n is in the center of G we see that $gd = \exp \sum_{i=1}^{n} b_i X_i$ and that d is the unique element of D for which $gd \in \exp F$. q.e.d

LEMMA. Let $M = G/D$ be as in the previous lemma. For each positive integer k let

$$(U)^k = \{g^k | g \in U\} .$$

If U is an open subset of G then there is a k such that $(U)^k$ contains a fundamental domain of M.

PROOF. We observe that $\log (U)^k = k(\log U)$. Since U is open it follows that $\log U$ is open. There is a k such that $k(\log U)$ contains a unit cube F. If follows that $\exp(F) \subset (U)^k$ and the previous lemma asserts that $\exp(F)$ is a fundamental domain. q.e.d.

§7. Distal and Pointwise Almost Periodic Flows.

THEOREM 3. Let G be a connected, simply-connected, nilpotent Lie group and let D be a uniform, discrete subgroup so that $M = G/D$ is a nilmanifold. Let φ be a one-parameter subgroup of G and let φ^* be the induced flow on M. Then φ^* is distal on M.

PROOF. Suppose φ^* is not distal on M. Then there are distinct points x and y in M and times t at which the orbits $\varphi_t^*(x)$ and $\varphi_t^*(y)$ approach arbitrarily closely. Using the compactness of $M = G/D$ there exists a point $z \in M$ and a sequence $\{t_m\}$ such that

$$\lim_{m\to\infty} \varphi_{t_m}^*(x) = z \quad \text{and} \quad \lim_{m\to\infty} \varphi_{t_m}^*(y) = z .$$

Now let

$$c : R^n \longrightarrow G$$

be a diffeomorphism of R^n onto G, carrying $\mathbf{Z}^n$ onto D, as determined by canonical coordinates. Let $\tilde{x} = (x_1,\ldots,x_n)$, $\tilde{y} = (y_1,\ldots,y_n)$, and project to x,y, and z, respectively. We can select $\tilde{x}$ and $\tilde{y}$ so that $|x_i - y_i| < 1$ for $i = 1,2,\ldots,n$.

Since the natural projection $G \longrightarrow G/D$ is a local homeomorphism, there exist sequences $d(m) = (d_1(m),\ldots,d_n(m))$ and $\delta(m) = (\delta_1(m),\ldots,\delta_n(m))$ in D with

$$\varphi(t_m)\,\tilde{x}\,d(m) = (x_1(m),\ldots,x_n(m)) \longrightarrow \tilde{z}$$

and

$$\varphi(t_m)\,\tilde{y}\,\delta(m) = (y_1(m),\ldots,y_n(m)) \longrightarrow \tilde{z}$$

as $m \longrightarrow \infty$.

Now

$$|x_1(m) - y_1(m)| = |x_1 - y_1 + d_1(m) - \delta_1(m)| .$$

However

$$\lim_{m\to\infty} |x_1(m) - y_1(m)| = 0$$

so $d_1(m) = \delta_1(m)$ and $x_1 = y_1$, for all large m.

Now suppose $d_i(m) = \delta_i(m)$ and $x_i = y_i$ for $i = 1,2,\ldots,j < n$. But it is easy to see that

$$|x_i(m) - y_i(m)| = |x_i - y_i + d_i(m) - \delta_i(m) + \theta_i|, \quad \text{for } i = 1,\ldots,n,$$

where θ_i are polynomials in the first $(i-1)$ canonical coordinates of $\tilde{x}$, $\tilde{y}$, $\varphi(t_m)$, $d(m)$, $\delta(m)$; and $\theta_i = 0$ when these coordinates of $\tilde{x}$ and $\tilde{y}$ and of $d(m)$ and $\delta(m)$ are the same. Thus

$$|x_{j+1}(m) - y_{j+1}(m)| = |x_{j+1} - y_{j+1} + d_{j+1}(m) - \delta_{j+1}(m)| .$$

Since

$$\lim_{m \to \infty} |x_{j+1}(m) - y_{j+1}(m)| = 0 ,$$

$x_{j+1} = y_{j+1}$ and $d_{j+1}(m) = \delta_{j+1}(m)$ for all large m. This shows that $\tilde{x} = \tilde{y}$, which contradicts the hypothesis that x and y are distinct in M. Therefore φ^* is distal on M.

COROLLARY. The discrete flow generated by $\varphi_1^* : x \longrightarrow \varphi_1^*(x)$, for $x \in M$, is distal on M.

If the induced flow φ^* on the nilmanifold M is not effective, then every point on M has a periodic orbit. If φ^* is effective, that is, only for $t = 0$ is φ_t^* the identity homeomorphism, then $\mathcal{G} = \{\varphi_t^* | -\infty < t < \infty\}$ is a group of homeomorphisms of M onto itself. Now topologize $\mathcal{G}$ by pointwise convergence, that is, consider the elements of $\mathcal{G}$ to be in the space M^M of (possibly discontinuous) transformations of M into M. Here

$$M^M = \prod_{x \in M} M_x, \quad \text{each } M_x = M,$$

with the product topology. Then M^M is a compact Hausdorff space. Consider the closure $\overline{\mathcal{G}}$ of $\mathcal{G}$ in M^M. By an extremely ingenious method, using the distal property of $\mathcal{G}$, R. Ellis [3] showed that $\overline{\mathcal{G}}$ is also a group of invertible transformations of M onto M. Now $\overline{\mathcal{G}}$ is not a topological group, but left translations by elements of $\mathcal{G}$, can be seen to give homeomorphisms of $\overline{\mathcal{G}}$ onto itself. Ellis then shows that there is a finite set $K \subset \mathcal{G}$ such that every element of $\mathcal{G}$ differs by an element of K from a transformation which maps a prescribed point $P \in M$ into a prescribed open neighborhood U of P. Thus, Ellis [1] shows that a distal group of homeo- of a compact Hausdorff space is pointwise almost periodic. The result holds for discrete flows as well as for continuous flows. We summarize this result in a theorem.

THEOREM 4. Let G be a connected, simply-connected, nilpotent Lie group and let D be a uniform discrete subgroup so that $M = G/D$ is a nilmanifold. Let φ be a one-parameter subgroup of G and let φ^* be the induced flow on M. Then φ^*, and also the discrete flow generated by φ_1^*, are pointwise almost periodic on M.

§8. Regionally Transitive and Minimal Flows. Consider an induced flow φ^* on a nilmanifold $M = G/D$, as above. Let one point $P \in M$ have a positive half-orbit, $\varphi_t^*(P)$ for $t \geq 0$, which is dense in M.

Then φ^* is regionally transitive on M; that is, for each pair of non-empty open subsets 0_1 and 0_2 in M, $\varphi_t^*(0_1)$ on $-\infty < t < \infty$ meets 0_2.

In case φ^* is regionally transitive on M, all points of M, excluding a set of first category, have positive and negative half-orbits which are dense in M. This is easily seen (Gottschalk, Hedlund [1]) since the set of points in M whose positive half-orbits fail to meet a given non-empty open set $0 \subset M$ is closed and without interior and hence nowhere dense. But there is a countable base for the open sets of M and hence the points of M whose positive half-orbits fail to meet even one non-empty open set of M is necessarily of first category. The negative half-orbits are treated similarly.

LEMMA. A regionally transitive flow which is pointwise almost periodic on a compact metric space X is minimal on X.

PROOF. (cf. Gottschalk, Hedlund [1].) Suppose $P \in X$ is a point whose orbit fails to meet a non-empty open set $0 \subset X$. Take some point $Q \in 0$ such that the orbit of Q is dense in X and so passes arbitrarily near to P.

Now the orbit of Q returns to a neighborhood $0_1 \subset 0$ of Q, with $\bar{0}_1 \subset 0$, at a relatively dense set of times. Thus the orbit of P must meet 0, which is a contradiction. Thus each point $P \in X$ has a dense orbit in X and so X is minimal. q.e.d.

THEOREM 5. Let G be a connected, simply-connected, nilpotent Lie group and D a uniform discrete subgroup so that $M = G/D$ is a nil-manifold. For each tangent vector v in L, the Lie algebra of G, there is a one-parameter subgroup $\varphi(t)$, with $\dot{\varphi}(0) = v$ in G, and an induced flow φ^* in M. Then there exists a subset E of first category in L such that every tangent vector of $L - E$ induces a flow on M which is regionally transitive and hence minimal.

PROOF. Let $C : R^n \longrightarrow G$ be a diffeomorphism of R^n onto G, carrying $\mathbf{Z}^n$ onto D, as determined by canonical coordinates. Let $P \in M$ be the image of $\tilde{P} = (0,0,\ldots,0)$ under the natural projection $G \longrightarrow G/D = M$.

Consider an open set $0_1 \subset M$. Let $E_1 \subset L$ be those initial tangent vectors for each of which the corresponding induced flow on M yields an orbit of P which fails to meet 0_1. Then E_1 is closed in L. Let $\mathring{E}_1$ be the interior of E_1. The orbits of P under the flows induced by vectors of $\mathring{E}_1$ must contain a non-empty open set $U \subset M$, if $\mathring{E}_1$ is non-empty. But, by the lemma in section 6, the powers $(U)^K$ must eventually contain a fundamental domain for M in G. Therefore the orbit of P meets 0_1, for some flow corresponding to a vector in $\mathring{E}_1$. Thus $\mathring{E}_1$ must be empty and E_1 is nowhere dense in L.

Take a countable base $0_1, 0_2, 0_3, \ldots$ for the open sets of M. For each $m = 1,2,3,\ldots$ the corresponding set $E_m \subset L$ yields all flows for which the orbit of P fails to meet 0_m. Then $E = \bigcup_{m=1}^{\infty} E_m$ is a first category set in L and E contains all vectors whose corresponding induced flows on M yield an orbit for P that is not dense in M.

Thus each vector in L - E defines a regional transitive flow and hence a minimal flow on M. q.e.d.

COROLLARY. There exists a first category set $\hat{E} \subset L$ such that for each induced flow φ^* on M corresponding to a vector in $L - \hat{E}$ and for each rational $c \neq 0$ the discrete flow on M generated by

$$\varphi_c^* : x \longrightarrow \varphi_c^*(x)$$

is minimal.

PROOF. Certainly φ_c^* is distal and pointwise almost periodic, as stated above. For each fixed rational $c \neq 0$ define E_{cm} to be those vectors producing discrete flows φ_c^* for which the orbit of P fails to meet the open set $O_m \subset M$, $m = 1, 2, 3, \ldots$. Then each E_{cm} is closed in L and has empty interior. Thus $E_c = \bigcup_{m=1}^{\infty} E_{cm}$ is of first category in L and contains all vectors for which the corresponding discrete flows φ_c^* yield a non-dense orbit of P.

Thus for each vector in $L - E_c$ the discrete flow φ_c^* is regionally transitive on M. But a regionally transitive, pointwise almost periodic, discrete flow on M is minimal; as is shown by an argument similar to that for the continuous flow.

Finally define $\hat{E} = \bigcup E_c$ for all rationals $c \neq 0$. Then $\hat{E}$ is of first category in L and is the required set. q.e.d.

THEOREM 6. Let G be a connected, simply-connected, nilpotent Lie group and let D be a uniform, discrete subgroup so that $M = G/D$ is a nilmanifold of dimension $n \geq 2$. Then there exists a non-countable number of topologically distinct flows on M, each of which is a minimal flow on M.

PROOF. Consider the induced flows on M corresponding to one-parameter subgroups of G. There is a set E of first category in the tangent space L at the identity of G such that vectors in L - E yield minimal flows on M.

Now for each induced flow φ^* on M, corresponding to the vector $v \in L - E$ the set of vectors of L, which define flows on M topologically equivalent to φ^*, has been proved to be of first category in L. If there were only a finite or countable collection of topological classes of minimal flows, then L - E would be the countable union of first category sets, which is impossible. Thus there are a non-countable number of topological equivalence classes of minimal flows on M. q.e.d.

REMARK. If the nilmanifold M is not a torus, then each minimal flow on M is not equicontinuous.

CHAPTER V.

NILFLOWS, MEASURE THEORY

by

L. Green

§1. The existence of minimal, distal, nonequicontinuous flows on compact nilmanifolds was proved in Chapter IV. The proof was nonconstructive, and gave no information on the measure-theoretic behaviour of these flows. As explained in Chapter II; however, these flows should be particularly susceptible to an attack by group representation theory. This is, in fact, the case. In the present chapter we give a complete qualitative description of the spectra of these flows. As a result we can tell precisely which one-parameter subgroups give rise to ergodic flows, and these are also the minimal flows. This gives an alternate proof to some of the results of Chapter IV.

In §2 we prove a lemma on induced representations needed later. §3 describes the groups we consider and contains the basic theorem about their irreducible unitary representations. These results are applied to flows on nilmanifolds in §4.

All the groups considered in this chapter are unimodular. Consequently, their homogeneous spaces by closed subgroups will have naturally defined invariant measures. Unless the contrary is expressly stated, it should be assumed that these measures are being used.

§2. Let D and H be closed subgroups of the locally compact, separable group G, such that H is normal and DH is closed. Suppose further that DH/D has finite volume. Denote the natural homomorphism of G on G/H by φ and set $D' = \varphi(D)$. Suppose L is a unitary representation of D which is the identity on $D \cap H$. Then $L = L' \circ \varphi$, where L' is a representation of D' in the same Hilbert space $\mathfrak{h}(L)$.

LEMMA 2.1. There exists a representation W of G such that

(1) $$U^L = W \oplus (U^{L'} \circ \varphi) .$$

Moreover, every subrepresentation of W is non-trivial on H.

PROOF. Let μ_1, μ_2, μ_3 be the invariant measures (induced by the corresponding Haar measures) for the homogeneous spaces G/D, G/DH, and DH/D, respectively. For $x \in G$, $y \in DH$ set $\alpha(x) = Dx$, $\beta(x) = DHx$,

$\gamma(y) = Dy$, considered as elements in G/D, G/DH, and DH/D, respectively. Since $G \supset DH \supset D$, we have for any $F \in L_2(G/D)$,

$$\int_{G/D} F(\alpha(x))\, d\mu_1(\alpha(x)) = \int_{G/DH} d\mu_2(\beta(x)) \int_{DH/D} F(\alpha(yx))\, d\mu_3(\gamma(y)).$$

(for a proof, see Mackey [2], Theorem 4.1, or Bruhat [1], p. 136). Identify the space G/DH with $(G/H)/D'$ by the canonical isomorphism.

Let $\|\cdot\|_L$ be the norm in $\mathfrak{h}(L)$, the representation space for L. $\mathfrak{h}(U^L)$ consists of all weakly measurable functions f from G to $\mathfrak{h}(L)$ such that $f(dx) = L_d f(x)$, $d \in D$, $x \in G$, with

$$\|f\|_1^2 = \int_{G/D} \|f(x)\|_L^2\, d\mu_1(\alpha(x)) < \infty .$$

Similarly, $\mathfrak{h}(U^{L'})$ consists of weakly measurable functions f' from G/H to $\mathfrak{h}(L)$ satisfying $f'(d'x') = L'_{d'} f'(x')$, and

$$\|f'\|_2^2 = \int_{G/DH} \|f'(x')\|_L^2\, d\mu_2(\beta(x)) < \infty ,$$

where $\varphi(x) = x'$. Consider the closed linear manifold $\mathfrak{M}$ of $\mathfrak{h}(U^L)$ consisting of those functions invariant under U_h^L, for all $h \in H$. Because H is normal, $\mathfrak{M}$ reduces $U^L : U^L = W \oplus W'$, where W is the restriction of U^L to $\mathfrak{M}^\perp$ and W' is the restriction to $\mathfrak{M}$.† We define a map V of $\mathfrak{M}$ into $\mathfrak{h}(U^{L'})$ by the equation

$$(Vf)(\varphi(x)) = \omega f(x) ,$$

where $\omega = \mu_3(DH/D)^{\frac{1}{2}}$. That the function on the right satisfies the first two conditions for being in $\mathfrak{h}(U^{L'})$ is immediate; we show that its norm is finite by showing that V is an isometry.

$$\|f\|_1^2 = \int_{G/D} \|f(x)\|_L^2\, d\mu_1(\alpha(x)) = \int_{G/DH} d\mu_2(\beta(x)) \int_{DH/D} \|f(hx)\|_L^2\, d\mu_3(\gamma(h))$$

$$= \int_{G/DH} \omega^2 \|f(x)\|_L^2\, d\mu_2(\beta(x)) = \|Vf\|_2^2 .$$

(Recall that $\|f(x)\|_L$ is a function on G/D, which, because $f(hx) = f(x)$ for $h \in H$, may also be considered as a function on G/DH.) Clearly $(Vf) \circ \varphi = \omega f$ and $V(F \circ \varphi) = \omega F$, for $F \in \mathfrak{h}(U^{L'})$, so V establishes an isometry between $\mathfrak{M}$ and $\mathfrak{h}(U^{L'})$.

---

† This decomposition exists for any unitary U and arbitrary-closed or not-normal H, provided one admits the cases when $\mathfrak{M}$ or $\mathfrak{M}^\perp$ can be (0). Moreover, W and W' are disjoint (vide infra).

Now

$$(U_g^{L'}Vf)(\varphi(x)) = (Vf)(\varphi(x)\varphi(g)) = (Vf)(\varphi(xg))$$
$$= \omega f(xg) = \omega(U_g^L f)(x) = (VU_g^L f)(\varphi(x)) ,$$

for $f \in \mathfrak{M}$. Hence $W' \cong U^{L'} \circ \varphi$, and by abuse of language we may write (1).

To conclude the proof it is only necessary to point out that W' was defined as the maximal subrepresentation of U^L which reduces to the identity on H.

COROLLARY 2.2. Suppose $U^L = \int_S M^s \, d\mu$ is a direct integral decomposition of U^L into irreducible parts, and $M_h^s = I$ for all $h \in H$ and $s \in S_H$, a subset of S. Then for almost all $s \in S_H$, there exist (irreducible) components V^s of a direct integral for $U^{L'}$ such that $M^s \cong V^s \circ \varphi$.

We remark that Lemma 2.1 holds with obvious modifications of the proof if the measures $\{\mu_i\}$ are merely quasi-invariant. Notice, also, that if $D \supset H$ the subspace $\mathfrak{M}$ is the whole space and the lemma becomes the familiar statement about how induced representations behave under homomorphisms. (Lemma 4.1, Mackey [3].)

§3. Let D be a discrete uniform subgroup of the connected, simply-connected nilpotent Lie group G. In Chapter IV a Malcev basis $\{X_1,\ldots,X_n\}$ of the Lie algebra $\mathfrak{G}$ of G which is adapted to D was described. We recall the following properties of such a basis:

i) If $\mathfrak{C}^k$ is the subspace of $\mathfrak{G}$ spanned by $\{X_k,X_{k+1},\ldots,X_n\}$, then $[\mathfrak{G},\mathfrak{C}^k] \subseteq \mathfrak{C}^{k+1}$ for $k = 1,2,\ldots,n-1$. (Thus the $\mathfrak{C}^k$ form a refinement of the descending central series.) In particular, each $\mathfrak{C}^k$ is an ideal of $\mathfrak{G}$.

ii) If $C^k = \exp \mathfrak{C}^k$, then $D \cap C^k$ is uniform in C^k.

iii) For some r, $\mathfrak{C}^{r+1} = [\mathfrak{G},\mathfrak{G}]$.

iv) Every element of D is uniquely expressible as a product

$$\exp(m_1X_1)\exp(m_2X_2)\ldots\exp(m_nX_n) ,$$

where $m_1,\ldots,m_n$ are integers.

v) The structure constants of $\mathfrak{G}$ with respect to this basis are rational numbers.

A subspace complementary to $[\mathfrak{G},\mathfrak{G}]$ is spanned by $X_1,\ldots,X_r$, and every homomorphism ρ of G into the reals R is given by a linear form on these r vectors. We call ρ <u>rational with respect to the given Malcev basis</u> (or merely <u>rational</u>, if there is no danger of confusion) if the scale in R may be chosen relative to which the coefficients of this form in the $\{X_i\}$ are rational.

A nilmanifold G/D has a natural measure invariant under the action of G, and $(U_g f)(Dx) = f(Dxg)$ defines a unitary representation of G in $L_2(G/D)$. This is precisely U^L, where L is the identity representation I_D, of D.

We let $L_2(R;\mathfrak{h})$ denote the space of norm-square-integrable measurable functions from the reals to the Hilbert space $\mathfrak{h}$, Lebesgue measure being understood.

THEOREM 3.1. Let V be an irreducible component of U^{I_D} which is not one-dimensional. Then (within equivalence) $\mathfrak{h}(V) = L_2(R;\mathfrak{h})$ for some Hilbert space $\mathfrak{h}$ and

$$(1) \qquad (V_g f)(x) = \mathfrak{A}(g,x) f(x + \rho(g)) \quad ;$$

here $\mathfrak{A}$ is a unitary operator on $\mathfrak{h}$ and ρ is a homomorphism of G into R, rational with respect to a given Malcev basis.

PROOF. We follow the proof of Lemma 12 in Dixmier [1], taking care now to preserve the rationality of ρ in the induction steps. If the dimension of G is one, the Theorem is vacuously true. Let us assume, then, that the theorem has been proved for all groups of dimension $< n$ which admit discrete uniform subgroups, and suppose that $\dim G = n$. We fix a Malcev basis of G with respect to D, with the properties described in the preceding paragraph, and let V be an irreducible component of U^{I_D}, non-trivial on $[G,G]$, (i.e., not one-dimensional).

$\mathfrak{C}^{n-1}$ is an Abelian ideal of $\mathfrak{G}$ of dimension 2, with basis $\{X_{n-1}, X_n\}$. For any $Y \in \mathfrak{G}$, $[X_{n-1}, Y] = \lambda(Y) X_n$ and $[X_n, Y] = 0$ (since $\mathfrak{C}^n$ is central). λ is a linear form on $\mathfrak{G}$ whose restriction to $\mathfrak{C}^{n-1}$ is zero. C^{n-1} is regularly embedded in G (Takenouchi [1], §4) and V restricted to C^{n-1} contains a character χ_0 such that V is equivalent to a representation induced by an irreducible representation of the stabilizer subgroup of χ_0. Let $\chi(\exp(t_{n-1}X_{n-1}) \exp t_n X_n) = \exp 2\pi i (t_{n-1}\chi^1 + t_n \chi^2)$ for any character χ of C^{n-1}. Call (χ^1, χ^2) the coordinates of χ (relative to the basis X_{n-1}, X_n). The orbit of χ under G has coordinates $(\chi^1 + \lambda(Y)\chi^2, \chi^2)$, as Y ranges over $\mathfrak{G}$. The stabilizer of χ_0 then is a connected subgroup G' whose Lie algebra $\mathfrak{G}'$ consists of all vectors $Y \in \mathfrak{G}$ such that $\lambda(Y)\chi_0^2 = 0$. Continuing to follow Dixmier's argument, we distinguish three cases:

i) $\chi_0^2 = 0$,

ii) $\lambda \equiv 0$,

iii) $\chi_0^2 \neq 0$, $\lambda \not\equiv 0$.

CASE i): If $\chi_0^2 = 0$, V restricted to C^n is the identity. Applying the lemma, we find that V is essentially an irreducible component of $U^{L'}$, where L' is the identity representation of $D' = D/D \cap C^n$. But

D' is a discrete uniform subgroup of G/C^n, and the images of the Malcev basis of $\mathfrak{G}$ form a Malcev basis of $\mathfrak{G}/\mathfrak{C}^n$, relative to D'. Since the image of the commutator is the commutator of the image, the homomorphism ρ' of G/C^n will be rational with respect to these images by the induction hypothesis. Hence ρ, which is the natural map onto the quotient followed by ρ', will also be rational.

CASE ii): If $x_0^2 \neq 0$, but $\lambda \equiv 0$, then C^{n-1} is central in G and V restricted to C^{n-1} must equal x_0. But V restricted to $D \cap C^{n-1}$ must be the identity, since that is the case for $U^{I}D$. Hence the coordinates of x_0 are integers, (m_1, m_2), and $m_2 \neq 0$. Let H denote the subgroup $\exp tY$, where $Y = X_{n-1} - \frac{m_1}{m_2} X_n$. V restricted to H is the identity, and $D \cdot H$ is closed. (To prove the latter, it is sufficient to show that $DH \cap C^{n-1}$ is closed, and this is merely a subgroup of the plane generated by an integral lattice and a line of rational slope. (See also Lemma 5, Matsushima [1].)) Thus the lemma applies again, and we find that the homomorphism ρ is rational.

CASE iii): Since $\lambda \not\equiv 0$ and $x_0^2 \neq 0$, the algebra $\mathfrak{G}'$ of the stabilizer subgroup consists of all vectors Y such that $\lambda(Y) = 0$. The homomorphism ρ of G onto G/G' is given by the formula $\rho(\exp X) = \lambda(X)$, provided we write G/G' as the additive group of reals. V is equivalent to the representation induced by an irreducible representation V' of G', and may be written as

$$(V_g f)(x) = \mathfrak{A}(g,x) f(x + \rho(g)) \quad ,$$

where f is in $L_2(R; \mathfrak{h}(V'))$. Now in terms of the Malcev basis, λ (and hence ρ) has rational coefficients, since the structure constants of $\mathfrak{G}$ in this basis are rational. This completes the proof of the theorem.

$L_2(R^p)$ will denote the Hilbert space of square-integrable functions on p-dimensional Euclidean space, R^p, with respect to Lebesgue measure.

COROLLARY 3.2: Let U be an irreducible representation of the connected nilpotent Lie group G which is not one-dimensional. Then there exists an integer $p \geq 1$, a real analytic function $\alpha(g, \vec{x})$ of absolute value one, and a homomorphism β of G into R^p such that (within equivalence) $\mathfrak{h}(U) = L_2(R^p)$ and

$$(U_g f)(\vec{x}) = \alpha(g, \vec{x}) f(\vec{x} + \beta(g)) \quad . \tag{2}$$

Furthermore, if U is a component of $U^{I}D$, where D is a discrete uniform subgroup of G, and G is now assumed to be simply-connected, then β can be chosen so that the p-th coordinate of $\beta(\exp Y)$ is rational with respect to a given Malcev basis.

PROOF: The first statement of this corollary is essentially found in Dixmier [1], and may be proved by induction by the method of

Theorem 3.1, where now D may be ignored. To obtain the sharpened form when U is an irreducible component of U^{ID}, we examine the proof of the theorem in a little more detail. Proceeding by induction on the dimension of G, we find that only case iii) must be reconsidered. In the notation used there, U is equivalent to a representation induced by an irreducible representation V' of a connected (simply-connected) closed subgroup G' of G. G' is nilpotent, so V' is a representation of the form (2), by the first part of this corollary. The induced representation acts in $L_2(R;L_2(R^p))$, which we identify in a natural way with $L_2(R^{p+1})$. The $(p+1)$-coordinate of $\beta(g)$ is precisely $\rho(g)$. The form of the multiplier is easily obtained, since G is a semi-direct product of G' and the reals. This completes the outline of the proof of the corollary.

We remark for later reference that $\alpha(e,\vec{x}) = 1$ for all $\vec{x} \in R^p$, since U is a representation.

§4. We write the elements of the compact nilmanifold $M = G/D$ as right cosets $\{Dg\}$ and denote the right translation action of G on M by φ_g; i.e., $\varphi_g(Dg') = Dg'g$. If K is the commutator subgroup of G, the projection $\Psi(Dg) = DKg$ fibers M over the torus $T = G/DK$. Setting $\psi_g(DKg') = DKg'g$, then $\Psi\varphi_g = \psi_g\Psi$.

With respect to the measures induced naturally from Haar measure on G, $\{\varphi_g\}$ and $\{\psi_g\}$ are groups of measure-preserving homeomorphisms on M and T, respectively. Consequently the equation $U_g f = f \cdot \varphi_g$ defines a unitary representation of G in $L_2(M)$; this is the representation induced by the identity representation of D. Since G acts transitively on M, the only functions invariant under all U_g are constants; in group-theoretic language, the identity representation of G is contained with multiplicity one in the representation U. When the expression for g is unwieldy as a subscript, we will write $U(g)$ for U_g.

According to the Mautner decomposition theorem, there exists a direct integral $\int \oplus \mathfrak{h}_s = L_2(M)$ and unitary representations U^s of G in $\mathfrak{h}_s$ such that $(U_g f)_s = U_g{}^s f_s$ and U^s is irreducible for almost all s. (Without danger of confusion, we denote by f_s the value at s of the image of $f \in L_2(M)$ under a fixed isomorphism with the direct integral space.) Let S_0 be the set of s such that $\mathfrak{h}_s$ is one-dimensional.

LEMMA 4.1. If $f_s = 0$ for $s \notin S_0$, then there exists a function $F \in L_2(T)$ such that $f(m) = F(\Psi(m))$ almost everywhere.

PROOF: Since U^s is irreducible, $U_g{}^s f_s = \alpha_s(g) f_s$ for $s \in S_0$, where α_s is a homomorphism of G into the complex numbers of modulus one. If $g \in K$, the commutator subgroup, $U_g{}^s f_s = f_s$, and by the hypothesis on f, this last equation holds for all s. Hence $f(\varphi_g(m)) = f(m)$ almost everywhere for $g \in K$. But K is transitive on the fibers of Ψ, so f is constant on these fibers.

Now let $\{g_t\}$ be a one-parameter subgroup of G with infinitesimal generator X, i.e., $g_t = \exp tX$. Set $\varphi_t = \varphi_{g_t}$ (and $\psi_t = \psi_{g_t}$) and call the resulting dynamical system $(M,\{\varphi_t\})$ a <u>nilflow</u>. Since we are dealing with a finite measure space, the assertion that $(M,\{\varphi_t\})$ is ergodic is equivalent to the assertion that $U(g_t)$ leaves only the constant functions invariant.

THEOREM 4.2. The nilflow $(M,\{\varphi_t\})$ is ergodic if and only if the torus flow $(T,\{\psi_t\})$ is ergodic. In the ergodic case, every eigenfunction of $U(g_t)$ is equivalent to a function $\chi \cdot \Psi$, where χ is a character of T. The remaining spectrum of $U(g_t)$ is Lebesgue.

PROOF. If $(T,\{\psi_t\})$ is not ergodic, there exists a non-constant invariant function F in $L_2(T)$. Then $F \cdot \Psi$ is a non-constant function in $L_2(M)$, invariant under $\{\varphi_t\}$. Hence we may assume for the remainder of the proof that $(T,\{\psi_t\})$ is ergodic. But all such flows on the torus are known: in terms of the Malcev basis of $\mathfrak{G}$, $\{\psi_{g_t}\}$ induces an ergodic flow on T if and only if X, the infinitesimal generator of $\{g_t\}$, has as components with respect to $\{X_1,X_2,\ldots,X_r\}$ a rationally independent set of numbers. Suppose $f \in L_2(M)$ is an eigenfunction for $U(g_t)$; i.e., $U(g_t)f = e^{i\lambda t}f$, where $\lambda = 0$ is a possibility. Then $U^s(g_t)f_s = e^{i\lambda t}f_s$ for almost all s. But if $s \notin S_0$, U^s is equivalent to a representation of the form (1) of Theorem 3.1. That is,

$$e^{i\lambda t}f_s(x) = \mathfrak{A}_s(g_t,x)f_s(x + \rho_s(g_t)) \quad ,$$

which implies, since $\mathfrak{A}_s$ is unitary, that

$$\|f_s(x)\| = \|f_s(x + \rho_s(\exp tX))\|$$

for all t and almost all x (the norm is taken in the range of f). Now $\rho_s(\exp tX) = t\rho_s(\exp X)$, and $\rho_s(\exp X) \neq 0$ because ρ_s is rational with respect to the same basis in which X has rationally independent components. Hence $\|f_s(x)\|$ can belong to $L_2(R)$ if and only if $f_s = 0$. Applying Lemma 4.1, we conclude that $f = F \cdot \Psi$, and F is an eigenfunction of $(T,\{\psi_t\})$, with the same eigenvalue. In particular, $\lambda = 0$ is impossible, so $(M,\{\varphi_t\})$ is ergodic. The eigenfunctions of $(T,\{\psi_t\})$ are precisely the characters of T as an abelian group.

To show that the remaining spectrum of $U(g_t)$ is Lebesgue, we apply Proposition 3.2. Each irreducible component of $U(g_t)$ which is not one-dimensional has as infinitesimal generator (recall that $\alpha(e,\vec{x}) = 1$)

$$\frac{1}{i}\,\beta(\exp X) \cdot \text{grad} + \frac{1}{i}\frac{d\alpha}{dt}\bigg|_{t=0} \cdot I \quad .$$

If $\beta(\exp X)$ is not the zero vector in R^p, a standard argument shows that this operator is unitary equivalent to $\frac{1}{i}\,\beta(\exp X) \cdot \text{grad}$. The latter has constant coefficients, so a Fourier transform shows that its spectrum is

Lebesgue. This completes the proof of the theorem.

In the following corollaries to Theorem 4.2, the standing assumptions are that $(M,\{\varphi_t\})$ is ergodic and M is not itself equal to T (i.e., G is not abelian).

COROLLARY 4.3. $(M,\{\varphi_t\})$ is not equicontinuous.

PROOF: Since $M \neq T$, there is some continuous spectrum. But the spectrum of an equicontinuous ergodic flow is pure point.

COROLLARY 4.4. The ergodic flows on M generated by independent one-parameter subgroups are metrically and topologically inequivalent.

PROOF: We say the one-parameter subgroups $\{\exp tX\}$ and $\{\exp t\tilde{X}\}$ are independent if there is no matrix (A_{ij}) with integer components and determinant plus or minus one such that

$$x_i = \sum_{j=1}^{r} A_{ij}\tilde{x}_j, \quad i = 1,\ldots,r,$$

where $\{x_i\}$, $\{\tilde{x}_i\}$ are the components of X and $\tilde{X}$ with respect to a given Malcev basis. By the theorem, the point spectrum of $U(\exp tX)$ is precisely the point spectrum of the flow induced on T, and this is the abelian group on r generators $\{\exp 2\pi i x_k\}$, $k = 1,\ldots,r$. These eigenvalues are metric (i.e., measure isomorphic) invariants, so $\{\exp t\tilde{X}\}$ yields the same eigenvalues if and only if it is dependent on $\{\exp tX\}$ in the above sense. In the present situation, since the eigenfunctions are continuous, these eigenvalues are also topological invariants of the flow.

COROLLARY 4.5. $(M,\{\varphi_t\})$ is a minimal flow.

PROOF: Since the measure of any open set of M is positive, ergodicity implies topological transitivity, which in turn implies that there exists a dense orbit. But it has been shown in the preceding chapter that the flow $(M,\{\varphi_t\})$ is distal. Hence, just as was pointed out there, minimality is a consequence of the existence of a dense orbit and Ellis' Theorem (Ellis [1]).

In conclusion we remark that the presence of mixed spectrum also shows that minimal nilflows are metrically distinct from the previously known minimal flows, namely, the flows on compact abelian groups with pure point spectrum and the horocycle flows with Lebesgue spectrum.

Corollary 4.5 has recently been strengthened by H. Furstenberg who proved (without using Ellis' theorem) that these flows are strictly ergodic. Meanwhile, Y. G. Sinai has found a complete set of metric invariants (the generalized eigen values) for ergodic nilflows.

CHAPTER VI.

FLOWS ON CERTAIN SOLVMANIFOLDS NOT OF TYPE E

by

L. Auslander and F. Hahn

§1. Introduction. In chapter III we examined the behaviour of the one-parameter flows $(S/D, \varphi(t))$. S was a connected, simply-connected, non-compact, non-nilpotent, three dimensional Lie group. D was a discrete subgroup such that S/D was compact and $\varphi : T \longrightarrow S$ was a one-parameter subgroup. In dimension three we found that there were two distinct types of such solvable groups S_1 and S_2 which had discrete subgroups D for which S_1/D was compact. For S_1 the exponential map was onto and $(S_1/D, \varphi(t))$ was never minimal regardless of the subgroup D. S_2 however, showed remarkably different behaviour. The exponential map is not onto S_2. If D is not abelian then $(S_2/D, \varphi(t))$ is never minimal. If D is abelian then we showed that $(S_2/D, \varphi(t))$ is isomorphic to a straight line flow on a three dimensional torus.

Our aim here is to study in greater generality the phenomenon which occurred in S_2. In the remainder of this chapter a solvable or nilpotent Lie group will always be considered to be connected, simply-connected, and non-compact. We let S be a solvable Lie group with a discrete subgroup D such that S/D is compact. We further assume that there is an isomorphism $D \longrightarrow N$, where N is a nilpotent Lie group and if D^* is the image of D then N/D^* is compact. Under these conditions we will show that S is not of type E. We will also show that for suitable one-parameter subgroups $\varphi : T \longrightarrow S$ the flow $(S/D, \varphi(t))$ is isomorphic to a flow $(N/D^*, x^*(t))$ where $x^* : T \longrightarrow N$ is a one-parameter subgroup.

§2. Two Preliminary Theorems. Two theorems are fundamental to the proofs of the statements in the introduction. The first is due to L. Auslander [2] and the second to Wang [1].

THEOREM 2.1. Let S be a solvable Lie group and D a discrete uniform subgroup such that S/D is compact. Further let there be an isomorphism $D \longrightarrow D^* \subset N$ where N is a nilpotent Lie group and N/D^* is compact. Under these conditions there is a torus group C of automorphisms of N and an isomorphism $S \longrightarrow C \cdot N$, where dot is the semi-direct product.

As is customary we will write $S \subset C \cdot N$ and $D \subset N$. The manner in which S lies in $C \cdot N$ is very important and examination of the proof of this theorem in Auslander [2] reveals the following information. Let $\pi : C \cdot N \longrightarrow N$ be the projection map. This induces a mapping $\pi^*: C \cdot N/D \longrightarrow N/D$ and π^* restricted to S/D is a homeomorphism of S/D onto N/D. This shows us that there are no non-trivial one-parameter subgroups of S which lie entirely in C. Suppose $\varphi : T \longrightarrow S$ is a one-parameter subgroup and $\varphi(t) = (c_t, e)$. We see that $\pi^*(c_t, e) = D$ and $\pi^*(e,e) = D$. Since π^* is 1:1 on S/D it follows that $c_t = e$ for all $t \in T$. We also learn that if $\nu : C \cdot N \longrightarrow C$ is the projection map then it induces a map $\nu^*: C \cdot N/D \longrightarrow C$ such that S/D is mapped onto C.

If H is the maximum normal analytic nilpotent subgroup of S then H is contained in N. The mapping $S/H \longrightarrow C$ induced by ν is an onto mapping. It is also known that N/H is abelian and that there is a mapping of this onto C.

If N is any nilpotent Lie group then the exponential map is a diffeomorphism of the Lie algebra of N onto N itself. We shall identify N with its Lie algebra by means of this mapping (see for instance Malcev [1] or Wang [1]). N thus has both a Lie group structure and a Lie algebra structure. These are related by

$$x \cdot y = x + y + [x,y]/2! + [x[x,y]]/3! + \ldots ,$$

where the right hand side is the Hausdorff-Baker-Campbell formula. Any automorphism of the group structure of N is an automorphism of the algebra N and conversely. We say an automorphsim of a nilpotent Lie group N is semi-simple if when it is considered as a linear transformation of the algebra structure of N it is completely reducible.

THEOREM 2.2. (Wang [1] p. 7.) If ρ is a semi-simple automorphism of a nilpotent Lie group N and $x \in N$ then there is a $y \in N$ such that $\rho(y)xy^{-1}$ is left fixed by ρ.

§3. The Extension of Wang's Theorem. In order to obtain the results of section 1 we must generalize Wang's theorem.

LEMMA 3.1. Let C be a compact group of automorphisms of a nilpotent Lie group N. Let N_1 be the linear subspace generated by the spaces $(I - c)N$, $c \in C$ and let N_2 be the set of fixed points of C. Under this condition $N = N_1 \oplus N_2$, N_1 is a subalgebra of N, and N_2 is an ideal in N.

PROOF: Since C is compact there is an inner product on the vector space structure of N such that C is a group of unitary operators. It thus follows that each element of C is semi-simple. Thus for each $c \in C$ we have $N = N_1(c) \oplus N_2(c)$ where $N_1(c) = (I - c)N$ and $N_2(c) =$ kernel $(I - c)$. We see that $N_2 = \bigcap_{c \in C} N_2(c)$ and N_1 is the space spanned by $N_1(c)$, $c \in C$. We thus see easily that $N = N_1 \oplus N_2$. Since each $N_1(c)$

is the image of N it is a subalgebra of N and thus N_1 is a subalgebra. Since each $N_2(c)$ is the kernel $(I - c)$ it is an ideal and thus N_2 is an ideal.

We add the following remarks to the above theorem. If $x,x' \in N_1$ and $y,y' \in N_2$ then $(x+y) \cdot (x'+y') = x \cdot x' + z$ where z is some element of N_2 and $x \cdot x' \in N_1$. An entirely analogous theorem is true if the compact group C is replaced by a single semi-simple automorphism of N.

THEOREM 3.2. Let $C \cdot N$ be the semi-direct product of a nilpotent Lie group N and a torus group C of automorphisms of N. If $\varphi(t) = (c_t, x_t)$ is a one-parameter subgroup of $C \cdot N$ then there is a $y \in N$ such that the one-parameter subgroup $\varphi^*(t) = y\varphi(t)y^{-1} = (c_t, x_t^*)$ has the property that c_t acts trivially on x_t^* for each $t \in T$.

PROOF. The idea of the proof is to use Wang's theorem to show that this is true for some real number s. We then show that the theorem is true for all reals of the form $m \cdot s/2^n$ where m and n are integers. Considerations of continuity complete the proof.

Choose $s > 0$ such that if $0 < |t| \leq s$ then $c_t + I$ is non-singular and $c_t - I$ has maximum rank. Since C is compact each element is semi-simple and thus by theorem 2.2 there is a $y \in N$ such that c_s acts trivially on $c_s(y)x_sy^{-1}$. We let $g^*(t) = y\varphi(t)y^{-1} = (c_t x_t^*)$ where $x_t^* = c_t(y)x_ty^{-1}$. Again we let $N_1(t) = (I - c_t)N$ and $N_2(t)$ be the fixed point set of c_t.

We will now show by induction that c_{ms} acts trivially on x_{ms}^* and $x_{ms}^* = (x_s^*)^m$ for each positive integer m. The statement has been established for $m = 1$. Suppose it is true for $m = p$. We then have

$$\begin{aligned}(c_{(p+1)s}, x^*_{(p+1)s}) &= \varphi^*((p+1)s) = (c_s, x_s^*) \cdot (c_{ps} \cdot x_{ps}^*) \\ &= (c_{(p+1)s}, c_{ps}(x_s^*)\, x_{ps}^*) = (c_{(p+1)s}, x_s^* \cdot x_{ps}^*) \\ &= (c_{(p+1)s}, (x_s^*)^{p+1}) .\end{aligned}$$

Therefore $(x_s^*)^{p+1} = x^*_{(p+1)s}$. Consequently we also have

$$c_{(p+1)s}\,(x^*_{(p+1)s}) = x^*_{(p+1)s} .$$

We now show that $c_{-s}(x^*_{-s}) = x^*_{-s}$. Since $\varphi^*(-s) = (\varphi^*(s))^{-1}$ we have

$$\begin{aligned}(e,e) &= (c_s, x_s^*)(c_{-s}, x^*_{-s}) = (e, c_{-s}(x_s^*) \cdot x^*_{-s}) \\ &= (e, c_s^{-1}(x_s^*) \cdot x^*_{-s}) = (e, x_s^* \cdot x^*_{-s}) .\end{aligned}$$

We therefore have $x^*_{-s} = (x_s^*)^{-1}$. Consequently $c_{-s}(x^*_{-s}) = x^*_{-s}$.

We now wish to show that $c_{s/2}(x^*_{s/2}) = x^*_{s/2}$. This is the same as proving that $x^*_{s/2} \in N_2(s/2)$. Since $N = N_1(s/2) \oplus N_2(s/2)$ we may write $x^*_{s/2} = y + z$ where $y \in N_1(s/2)$ and $z \in N_2(s/2)$. We see that

$$(c_s, x^*_s) = \varphi^*(s) = (\varphi^*(s/2))^2 = (c_{s/2}, x^*_{s/2})^2 =$$
$$= (c_s, (c_{s/2}(y) + z)(y + z)) = (c_s, (c_{s/2}(y) \cdot y + z')),$$

where z' is in $N_2(s/2)$. Thus $x^*_s = c_{s/2}(y) \cdot y + z'$. Since c_s leaves x^*_s fixed we know that x^*_s has no component in $N_1(s)$ and therefore $c_{s/2}(y) \cdot y = 0$. Thus $c_{s/2}(y) = y^{-1} = -y$ and this together with the fact that $I + c_{s/2}$ is non-singular implies that $y = 0$. Thus $x^*_{s/2} = z \in N_2(s/2)$.

The above arguments show that $c_r(x^*_r) = x^*_r$ if $r = ms/2^n$ where m and n are integers. Because of continuity of one-parameter subgroups the result follows.

COROLLARY 3.3. With the same notation and hypothesis of the previous theorem we have $c_t(x^*_r) = x^*_r$ for t and r real.

PROOF. There is an $s > 0$ such that if $0 < |t| \leq s$ then $N_2(t)$ is constant. Thus if $|t|$ and $|r| \leq s$ we have $c_t(x^*_r) = x^*_r$. Since any numbers r and s may be written $r = mp$ $s = nq$ where m and n are integers and $|p|$ and $|q| < s$ we have $c_t(x^*_r) = x^*_r$.

We remark that what we have just shown is that x^*_t is a one-parameter subgroup of N.

§4. The Major Theorems. Let $C \cdot N$ be the semi-direct product just as in the previous sections. Let $\nu : C \cdot N \longrightarrow C$ be the projection and let $\varphi : T \longrightarrow C \cdot N$ be a one-parameter subgroup. We say that φ has maximal rank provided that $\nu \circ \varphi(T)$ does not lie in a torus whose dimension is lower than the dimension of C. We say that φ is in general position provided $\nu \circ \varphi(T)$ is dense in C. We remark that general position implies maximal rank but not conversely.

For the remainder of this section we let S be a solvable Lie group with a discrete subgroup D for which S/D is compact. We further assume that there is an isomorphism $D \longrightarrow N$, where N is a nilpotent Lie group and N/image (D) is compact.

THEOREM 4.1. If the flow $(S/D, \varphi(t))$ has a dense orbit then $\varphi(t)$ is in general position.

PROOF. Let $S \subset C \cdot N$ and let $\nu : C \cdot N \longrightarrow C$ be the projection. This projection defines a continuous map ν^* of $C \cdot N/D \longrightarrow C$. The restriction of ν^* to S/D defines a homomorphism of $(S/D, \varphi(t))$ onto $(C, \nu \circ \varphi(t))$. Since the first flow has a dense orbit so does the second flow. The second flow is a flow on a torus induced by a one-parameter subgroup $\nu \circ \varphi$ and thus all the orbits are dense. Consequently $\nu \circ \varphi(T)$ is dense in C.

We remark that if the flow $(S/D, \varphi(t))$ is ergodic or minimal then φ is in general position.

THEOREM 4.2. Let $S \subset C \cdot N$ and let $\varphi(t) = (c_t, x_t)$ be a one-parameter subgroup of S which is in general position. Under these conditions there is an isomorphism of the flow $(S/D, \varphi(t))$ onto the flow $(N/D^*, x_t^*)$ where $D^* \subset N$ is an isomorphic image of D and x_t^* is a one-parameter subgroup of N.

PROOF. By Corollary 3.3 there is an inner automorphism of $C \cdot N$ induced by an element $y \in N$ such that if $y\varphi(t)y^{-1} = \varphi^*(t) = (c_t, x_t^*)$ then x_t^* is a one-parameter subgroup of N. Since $y \in N$ this automorphism maps N onto itself and we let $D^* = yDy^{-1}$. This induces an isomorphism of the flow $(S/D, \varphi(t))$ onto the flow $(S/D^*, \varphi^*(t))$.

Let $\pi : C \cdot N \longrightarrow N$ be the projection map. This induces a map $\pi^*: C \cdot N/D^* \longrightarrow N/D^*$ and the restriction of π^* to S/D^* is a homeomorphism onto N/D^*. By Corollary 3.2 c_t acts trivially on x_r^* for each real t and r. Since φ is in general position we obtain that c acts trivially on x_r^* for each $c \in C$ and real r. We observe that $\pi^*(((c,x)D^*)_t) = \pi^*((c_t, x_t^*)(c,x)D) = \pi^*((c_t c, x_t^* x)D) = x_t^* xD = (xD^*)_t$. Thus π^* is equivariant and the theorem is complete.

We add that the questions of minimality and ergodicity of nilflows are completely worked out in Chapters IV and V.

THEOREM 4.3. If S is not nilpotent then S is not of type E.

PROOF. Let $S \subset C \cdot N$ as before. It follows from the proof of Theorem 1, Auslander [2] that $N = H \oplus N'$ where H is the maximum normal analytic nilpotent subgroup of S. There is a homomorphism $x \longrightarrow c_x$ of N' onto C such that the elements of S are precisely those elements of $C \cdot N$ of the form $(c_x, y + x)$ where $y \in H$ and $x \in N'$ are arbitrary.

By Lemma 2.1 we may also write $N = N_1 \oplus N_2$ where N_2 is the fixed point set of C and $N_1 \subset H$. Furthermore N_1 decomposes into two dimensional invariant subgroups on each of which C acts as a rotation group. We choose $y \in H$ such that it has no non-zero components in the above decomposition of N_1. Choose $x \in N'$ such that $x \neq 0$ and $c_x = e$. We will show that there is no one-parameter subgroup of S through $(e, y + x)$.

Suppose there is a subgroup $\varphi : T \longrightarrow S$ such that $\varphi(1) = (e, y + x)$. There is a two dimensional invariant subspace N'' of N_1 in which y has a non-zero component and φ acts non-trivially.

If we restrict φ to N'' then it is isomorphic to a matrix group

$$\begin{pmatrix} \cos 2\pi nt & \sin 2\pi nt \\ -\sin 2\pi nt & \cos 2\pi nt \end{pmatrix} \quad \text{where } n$$

is an integer. If $t = 1/2n$ we see that $(\varphi(1/2n))^{2n} = \varphi(1) = (e, y + z)$.

If we write $\varphi(t) = (c_t, x_t)$ we note the even powers of $\left(c_{\frac{1}{2n}}, x_{\frac{1}{2n}}\right)$ have zero components in N''. Thus $\varphi(1) = \left(c_{\frac{1}{2n}}, x_{\frac{1}{2n}}\right)^{2n} = (e, y + z)$ has a zero component in N'' which contradicts the choice of y. Consequently S is not of type E.

CHAPTER VII

FLOWS IN TYPE (E) SOLVMANIFOLDS

by

L. Green

§1. A connected solvable Lie group G is said to be of type (E) if the exponential map is onto. As examples, we note that the group S_1 of Chapter III is of type (E), while S_2 in the same chapter is solvable, but not of type (E). M. Saito ([1]) and Dixmier ([2]) have investigated these groups, and O. Takenouchi ([1]) has indicated a method of obtaining all their irreducible unitary representations. For the purpose of examining the ergodic properties of flows on solvmanifolds of such groups, however, it is not necessary to know all the representations; we can use Mautner's method, the essence of which consists in looking at the representations of small subgroups which fill up the group in some sense. This enables us to reduce the problem to that of a nilflow on an associated nilmanifold, and these flows are treated in detail in Chapter IV. Let D be a discrete uniform subgroup of the type (E) group G. The final result is that a one-parameter subgroup of G acts ergodically on the solvmanifold G/D if and only if it acts ergodically on the nilmanifold G/DR, where R is the smallest connected normal subgroup of G such that G/R is nilpotent.

As in the nilpotent and semi-simple cases, a change in the uniform subgroup D does not make any essential difference in the answer. This is in contrast to the situation when G is not of type (E), where already in three dimensions the opposite is true (cf. Chapter III).

In §2 we describe the root structure of type (E) algebras and establish the notation used in the sequel. The results needed from representation theory are reviewed in §3. In §4 we prove the basic result about flows mentioned above.

§2. Let $\mathfrak{G}$ be the real Lie algebra of the connected, simply-connected, solvable Lie group G. An element $X \in \mathfrak{G}$ is called regular if the multiplicity of zero as an eigenvalue of ad X is less than or equal to that of ad Y for all $Y \in \mathfrak{G}$. Denote the complexification of $\mathfrak{G}$ by $\mathfrak{L}$, and consider $\mathfrak{G}$ as the set of real elements of $\mathfrak{L}$ (i.e., the elements fixed under the semi-morphism $X \longrightarrow \bar{X}$). Let X_0 be a fixed regular element of $\mathfrak{G}$; X_0 is also regular in $\mathfrak{L}$. Considered as acting in $\mathfrak{L}$, ad X_0 induces the decomposition of $\mathfrak{L}$ into primary components

$$\mathfrak{L} = \sum \mathfrak{L}_\omega \ ,$$

where $\mathfrak{L}_\omega$ is the subspace of $\mathfrak{L}$ on which (ad $X_0 - \omega I$) is nilpotent; in particular, $\mathfrak{L}_0$ is the Cartan subalgebra generated by X_0. The subspaces $\mathfrak{L}_\omega$ for $\omega \neq 0$ are not in general subalgebras, but we have the relation

(1) $$[\mathfrak{L}_\omega, \mathfrak{L}_{\omega'}] \subset \mathfrak{L}_\gamma \ ,$$

where $\mathfrak{L}_\gamma = \mathfrak{L}_{\omega+\omega'}$ if $\omega + \omega'$ is a root, and equals the zero subspace otherwise.

Let $\mathfrak{G}_\omega (= \mathfrak{G}_{\bar\omega})$ be the set of real vectors in $\mathfrak{L}_\omega + \mathfrak{L}_{\bar\omega}$. Then the subspaces $\mathfrak{G}_\omega$ are the primary components of ad X_0 in $\mathfrak{G}$ and from (1) we derive

(2) $$[\mathfrak{G}_\omega, \mathfrak{G}_\sigma] \subseteq \mathfrak{G}_{\omega+\sigma} + \mathfrak{G}_{\omega+\bar\sigma} \ ,$$

where it is understood that a subspace with a subscript which is not a root is zero.

Now we make the assumption that G is a type (E) group. Then (Saito II) none of the roots are pure imaginary and we may write them in order of increasing real parts:

$$\nu_k \leq \nu_{k-1} \leq \cdots \leq \nu_1 < 0 < \pi_1 \leq \pi_2 \leq \cdots \leq \pi_r \ ;$$

here ν_i denotes a root with negative real part, π_i one with positive real part; the order in which roots with the same real part appear is unimportant. Notice that by our hypothesis, zero occurs with multiplicity one. We assume in what follows that G is not nilpotent, so the roots are not all zero.

As immediate consequences of (2) we have the following statements:

i) $[\mathfrak{G}_0, \mathfrak{G}_\omega] \subset \mathfrak{G}_\omega$.

ii) $\sum_{i=j}^{r} \mathfrak{G}_{\pi_i}$ is a subalgebra for $j = 1,2,\ldots,r$.

ii)' $\sum_{i=j}^{k} \mathfrak{G}_{\nu_i}$ is a subalgebra for $j = 1,2,\ldots,k$.

iii) $\sum_{i=j+1}^{r} \mathfrak{G}_{\pi_i}$ is an ideal in $\sum_{i=j}^{r} \mathfrak{G}_{\pi_i}$, $j = 1,2,\ldots,r-1$.

iii)' $\sum_{i=j+1}^{k} \mathfrak{G}_{\nu_i}$ is an ideal in $\sum_{i=j}^{r} \mathfrak{G}_{\nu_i}$, $j = 1,2,\ldots,k-1$.

We also note for later reference that, since ad X_0 is non-singular on $\mathfrak{G}_\omega$, $\omega \neq 0$, this latter subspace is contained in the commutator, $[\mathfrak{G},\mathfrak{G}]$.

Two special type (E) Lie algebras will be of importance in the sequel. The algebra with basis $\{X,Y\}$ and relation $[X,Y] = Y$ will be called $\mathfrak{S}_1$. (This is the algebra of the real affine group on the line.) $\mathfrak{S}_2$ will denote the three dimensional algebra with basis $\{X,Y,Z\}$ and relations $[X,Y] = Y - \sigma Z$, $[X,Z] = \sigma Y + Z$, $[Y,Z] = 0$, where σ is real and different from zero. This is a typical algebra of a type (E) group having a single pair of complex conjugate roots. Notice that any regular element W of $\mathfrak{S}_i$, $i = 1$ or 2, is contained in a basis like that described above, with $W = X$. The representation theory of the corresponding simply-connected groups has been treated in Chapter II.†

LEMMA 2.1. Let $\mathfrak{C}$ be a solvable type (E) Lie algebra with regular element X. Suppose that ad X has nullity one and a single non-zero characteristic value ω (respectively, a single pair of conjugate roots, ω and $\bar{\omega}$). Then there exists a sequence of ideals

$$(0) = \mathfrak{A}_0 \subset \mathfrak{A}_1 \subset \ldots \subset \mathfrak{A}_{\ell-1} \subset \mathfrak{A}_\ell = [\mathfrak{C},\mathfrak{C}]$$

such that $(\{X\} + \mathfrak{A}_i)/\mathfrak{A}_{i-1}$, $i = 1,2,\ldots,\ell$, is isomorphic to $\mathfrak{S}_1$ or $\mathfrak{S}_2$, respectively, according to whether ω is real or complex. The image of X is regular in each of these factor algebras.

PROOF: The root space $\mathfrak{C}_\omega$ in the decomposition of $\mathfrak{C}$ relative to ad X must coincide with $[\mathfrak{C},\mathfrak{C}]$, since it is always contained in the commutator, and here has codimension one. The inclusion (2) implies that in this situation, $\mathfrak{C}_\omega$ is abelian. Hence the decomposition of $\mathfrak{C}_\omega$ into cyclic subspaces relative to ad X is actually a decomposition into ideals, so the ascending chain of ideals may be constructed from these subspaces in the obvious way without worrying about their product structure. Since the nullity of the image of ad X remains one, the image of X is regular.

THEOREM 2.2. Let $\mathfrak{G}$ be a solvable Lie algebra with regular element X_0. There exist two sequences of subalgebras of $\mathfrak{G}$: $\mathfrak{A}_1 \subset \mathfrak{A}_2 \subset \ldots \subset \mathfrak{A}_\ell$, $\mathfrak{B}_1 \subset \mathfrak{B}_2 \subset \ldots \subset \mathfrak{B}_m$, with the following properties

i) Each $\mathfrak{A}_i(\mathfrak{B}_i)$ is invariant under ad X_0.

ii) $\mathfrak{A}_{i-1}(\mathfrak{B}_{i-1})$ is an ideal in $\{X_0\} + \mathfrak{A}_i$ (Resp. $\{X_0\} + \mathfrak{B}_i$).

iii) $(\{X_0\} + \mathfrak{A}_i)/\mathfrak{A}_{i-1}((\{X_0\} + \mathfrak{B}_i)/\mathfrak{B}_{i-1})$ is isomorphic to either $\mathfrak{S}_1$ or $\mathfrak{S}_2$, with the image of X_0 regular.

iv) $\mathfrak{A}_\ell = \sum_{R(\omega)>0} \mathfrak{G}_\omega$ and $\mathfrak{B}_m = \sum_{R(\omega)<0} \mathfrak{G}_\omega$. (In particular, if $\mathfrak{G}$ is not nilpotent and has one root which is not pure imaginary, at least one of these sequences is not trivial.)

† It is interesting to note that precisely these algebras occur naturally in the discussion of non-compact semi-simple groups (cf. Segal and von Neumann [1]).

PROOF: If π_r is a root of greatest positive real part, $\mathfrak{G}_{\pi_r}$ is a subalgebra invariant under ad X_0 and $\{X_0\} + \mathfrak{G}_{\pi_r}$ is an algebra satisfying all the hypotheses of the lemma. $\mathfrak{A}_1, \mathfrak{A}_2, \ldots, \mathfrak{A}_{\ell_r}$ are constructed according to the lemma, with $\mathfrak{A}_{\ell_r} = \mathfrak{G}_{\pi_r}$. Suppose the sequence of subalgebras has been constructed up to $\mathfrak{A}_{\ell_k} = \sum_{j=k}^{r} \mathfrak{G}_{\pi_j}$, $k < \ell$. Then $\mathfrak{A}_{\ell_k}$ is an ideal in $\{X_0\} + \sum_{j=k-1}^{r} \mathfrak{G}_{\pi_j}$, and the lemma may be applied to the factor algebra, in which ad X_0 has the single non-zero root π_{k-1}. Thus, the construction may be continued until all the root spaces belonging to roots with positive real part are exhausted. The same procedure is used to construct the $\mathfrak{B}_i$.

§3. Let S_i, $i = 1,2$, denote the simply-connected, connected Lie groups with algebras $\mathfrak{S}_i$, $i = 1,2$. We recall the results of Chapter II pertaining to the representations of these groups. Let X_0 be a regular element of $\mathfrak{S}_i$. Then there exists a basis for $\mathfrak{S}_i$ containing X_0 which looks precisely like the basis defining $\mathfrak{S}_i$, and, consequently, in which X_0 plays the role of X. Combining the Mautner Lemma with Corollary D of Chapter II, we may state,

LEMMA 3.1. Let $g \longrightarrow U(g)$ be a unitary representation of S_i, $i = 1$ or 2, in a Hilbert space $\mathfrak{H}$. If $\varphi \in \mathfrak{H}$ is such that $U(\exp tX_0)\varphi = e^{i\lambda t}\varphi$ for all t and some fixed λ, then $U(\exp Y)\varphi = \varphi$ for all $Y \in [\mathfrak{S}_i, \mathfrak{S}_i]$.

We can now obtain the analogous theorem for type (E) groups. If $\mathfrak{G}$ is a solvable Lie algebra, set $\mathfrak{G}^\infty = \lim_k [\underbrace{\mathfrak{G}, [\mathfrak{G}, \ldots, [\mathfrak{G}, \mathfrak{G}]}_{k} \ldots]]$.

THEOREM 3.2. Let $g \longrightarrow U(g)$ be a unitary representation of the connected, simply-connected solvable Lie group G of type (E). Let X_0 be regular in the Lie algebra $\mathfrak{G}$ of G. If $\varphi \in \mathfrak{H}(U)$ is such that $U(\exp tX_0)\varphi = e^{i\lambda t}\varphi$ for all t, and a fixed λ, then $U(\exp Y)\varphi = \varphi$ for all $Y \in \mathfrak{G}^\infty$.

PROOF. We shall assume the sequences described in Theorem 2.2 have been constructed for $\mathfrak{G}$ relative to X_0 and use the notation of that theorem. Let $\mathfrak{H}'$ be the smallest closed linear manifold of $\mathfrak{H}$ containing the orbit of φ under U. U restricted to the subgroup corresponding to the subalgebra $\{X_0\} + \mathfrak{A}_1$ is a representation of a homomorphic (actually, isomorphic) image of S_i for $i = 1$ or 2. Hence Lemma 3.1 tells us that $U(\exp Y)\varphi = \varphi$ for $Y \in \mathfrak{A}_1$. Let $\mathfrak{H}^k$ be the smallest closed linear manifold containing the orbit of φ under U restricted to the subgroup corresponding to $\{X_0\} + \mathfrak{A}_k$, and suppose we have shown that $U(\exp Y)\varphi = \varphi$ for $Y \in \mathfrak{A}_{k-1}$. Because $\mathfrak{A}_{k-1}$ is an ideal in $\{X_0\} + \mathfrak{A}_k$, this representation on $\mathfrak{h}^k$ reduces to a representation of the factor group whose Lie algebra is $(\{X_0\} + \mathfrak{A}_k)/\mathfrak{A}_{k-1}$, namely a group of type S_i with regular element the image of X_0. Hence U is trivial on $\mathfrak{A}_k$, also. In a similar fashion, we find that $U(\exp Z)\varphi = \varphi$ for any Z in $\mathfrak{B}_m$. The set of Z's for which this is true is a subalgebra containing $\mathfrak{B}_m$ and $\mathfrak{A}_\ell$, so it certainly contains $\mathfrak{G}^\infty$. Notice that $\mathfrak{G}^\infty$ is independent of the regular element X_0; for this property, and the observation that $\mathfrak{G}^\infty$ is the smallest subalgebra containing all the non-zero root spaces, see Mostow [1]. Thus, the proof is complete.

We remark that Theorem 3.2 holds even if G is not of type (E), provided we replace $\mathfrak{G}^\infty$ by the smallest subalgebra containing the root spaces (now relative to X_0) belonging to roots with non-vanishing real parts. For Theorem 2.2 is true for any solvable algebra.

§4. In this section G will be a connected, simply-connected solvable Lie group of type (E), $\mathfrak{G}$ its Lie algebra, and R the connected normal subgroup of G corresponding to the ideal $\mathfrak{G}^\infty$ described in the last section. G/R is a connected, simply-connected nilpotent Lie group. According to the Theorem proved in the appendix to this chapter, if D is a discrete uniform subgroup of G, $D \cap R$ is uniform in R. Thus, DR is closed and G/DR is a compact nilmanifold. The same theorem tells us that $D[G,G]$ is closed. We have the following commutative diagram of groups and homogeneous spaces (all maps indicated are the natural ones):

$$\begin{array}{ccccc} G & \xrightarrow{\Phi} & G/R & \xrightarrow{\Phi'} & G/[G,G] \\ \downarrow & & \downarrow & & \downarrow \\ G/D & \xrightarrow{\Psi} & G/DR & \longrightarrow & G/D[G,G] \end{array} .$$

Set $\Theta = \Phi' \circ \Phi$. (Despite the above notation, we shall use right cosets to make the group representations simpler.)

G acts on G/D on the right by the formula $g : Dg' \longrightarrow Dg'g$. There exists a natural measure on G/D invariant under this action of G. Thus, the formula $(U_g f)(Dg') = f(Dg'g)$ defines a unitary representation of G in $L_2(G/D)$, where the latter space is constructed with respect to this measure. (This is the representation of G induced by the identity representation of D.) Every one-parameter subgroup $\{g_t\}$ of G is of the form $\{\exp tX\}$ for some $X \in \mathfrak{G}$. Each such group defines a flow on the solvmanifold G/D by restricting the action of G to the subgroup. $\Phi(g_t)$ is a one-parameter (possibly trivial) subgroup of G/R, and hence induces a nilflow on the nilmanifold G/DR. Without danger of confusion, we shall use the same symbols for the one-parameter subgroups and for the flows they generate.

THEOREM 4.1. If X is a regular element of $\mathfrak{G}$, the flow $\{\exp tX\}$ on G/D is ergodic if and only if $\{\Phi(\exp tX)\}$ is ergodic on G/DR. In the ergodic case, every eigenfunction of $U(\exp tX)$ is the lift of an eigenfunction of the flow $\Theta(\exp tX)$ on the torus $G/D[G,G]$.

PROOF. Suppose $h(Dg)$ is an eigenfunction for $U(\exp tX)$. By Theorem 3.1, $h(Dgr) = h(Dg)$ for all $r \in R$. Let $H(g)$ be the function of G, constant on right cosets of D, whose projection to G/D is h. Then, by the normality of R,

$$H(drg) = H(dgr') = H(dg) = H(g)$$

for all $d \in D$, $r \in R$. Thus, $h = F \cdot \Psi$, where F is an eigenfunction for the flow $\Phi(\exp tX)$ with the same eigenvalue. In particular, h is constant under the flow if and only if F is constant under the corresponding nilflow. This proves the first part of the theorem. The proof of the theorem is completed by applying the nilflow theorem (Chapter V) to $\Phi(\exp tX)$.

COROLLARY: If D is a closed uniform subgroup of the solvable, non-abelian, connected, simply-connected type (E) group G, there exist ergodic flows on G/D generated by one-parameter subgroups of G. These flows are neither equicontinuous nor mixing.

PROOF. The homomorphism of $\mathfrak{G}$ onto $\mathfrak{G}/[\mathfrak{G},\mathfrak{G}]$ is continuous and open in the ordinary topology of these real vector spaces. The regular elements of $\mathfrak{G}$ form an open set whose image is consequently open, hence contains elements of any dense set. But the vectors of $\mathfrak{G}/[\mathfrak{G},\mathfrak{G}]$ giving rise to ergodic (irrational) flows on the torus $G/D[G,G]$ are dense, so at least one is the image of a regular element of $\mathfrak{G}$. Any such regular pre-image then induces an ergodic flow on G/D. By the theorem, the eigenfunctions cannot be dense in $L_2(G/D)$, so the spectrum is not pure point and the flow is not equicontinuous. To prove that it is not mixing, it is sufficient to show that there is at least one eigenfunction other than the constants, for then the spectrum cannot be purely continuous. But the existence of such a function would follow from the fact that the torus $G/D[G,G]$ is non-trivial. If G is nilpotent, this torus is at least two-dimensional. If G is not nilpotent, $D[G,G] \subset DN$, where N is the maximal normal nilpotent subgroup of G. Since $DN \neq G$, the torus is non-trivial and there exists some point spectrum for the flow.

In three dimensions there is only one type (E) solvable group which admits a discrete uniform subgroup — this is the group S_1 in Chapter III. The regular elements of its Lie algebra are precisely the elements which induce non-trivial rotations on the associated torus; since the latter is a circle, every such one-parameter group of rotations is ergodic. In this case we were able to show that periodic orbits exist, so the flow on the solvmanifold is not minimal. It seems reasonable to conjecture: if a flow on a type (E) solvmanifold S is minimal, then S is a nilmanifold and the flow is equivalent to a nilflow.

CHAPTER VII

Appendix

L. Auslander

Let G be a connected, simply-connected solvable Lie group of type (E) with a discrete uniform subgroup D. If S and H are subsets of G, $[S,H]$ will denote the group generated by elements of the form $shs^{-1}h^{-1}$, $s \in S$, $h \in H$. Set $G^{\infty} = \lim_{k \to \infty} [G,\ldots,[G,[G,G]]\ldots]$. We want to prove that $G^{\infty} \cap D$ is uniform in G^{∞}.

LEMMA. Let $\mathfrak{G}^*$ be the set of X in $\mathfrak{G}$, the Lie algebra of G, such that $(\exp tX) \in D$ for some $t \neq 0$. Then $\mathfrak{G}^*$ is dense in $\mathfrak{G}$.

PROOF. Let $\phi : G \longrightarrow G/D$ be the natural projection. Every one-parameter subgroup of G is of the form $\exp tX$ for some $X \in \mathfrak{G}$. If $\phi(\exp t_1X) = \phi(\exp t_2X)$ for $t_1 \neq t_2$, then $\exp(t_1-t_2)X \in D$ and $X \in \mathfrak{G}^*$. Thus ϕ is one-to-one on every one-parameter subgroup of the form $\exp tY$, $y \notin \mathfrak{G}^*$. Suppose Y is contained in an open subset (and hence an open cone) K of $\mathfrak{G} - \mathfrak{G}^*$. Because G is simply-connected and of type (E), the exponential map is a homeomorphism. Since ϕ is open, its restriction to $\exp K$ is a homeomorphism. Thus $\phi(\exp tY)$, $t > 0$, is a homeomorphic image of an infinite ray in $\mathfrak{G}$. This contradiction of the compactness of G/D proves the lemma.

THEOREM: Let G be a connected, simply-connected solvable Lie group of type (E) with a discrete uniform subgroup D. Let R be a connected normal subgroup such that $R/R \cap D$ is compact. Then $[G,R]/D \cap [G,R]$ is compact.

PROOF. $R \cap D$ is a discrete uniform subgroup of the solvable group R. If R is nilpotent, then $D \cap [R,R]$ is uniform in $[R,R]$. If R is not nilpotent, N, its maximal analytic nilpotent normal subgroup, has $N \cap D$ as a uniform subgroup (Mostow [1]). In any case, R contains a closed connected characteristic subgroup H in which $D \cap H$ is uniform and such that R/H is abelian. Thus $[G,R]/D \cap [G,R]$ is fibered over $[G/H,R/H]/(DH/H) \cap [G/H,R/H]$ with compact fiber $H/D \cap H$, so the compactness of the space in question would follow from the compactness of the base space. In other words, we may assume that R is abelian.

The element srs^{-1} may be written as $\mathrm{ad}(s)r$, where $\mathrm{ad}(s)$ is the corresponding linear transformation of the vector space R. Then $[G,R]$

is the subspace spanned by $(\mathrm{ad}(g) - I)R$ as g ranges over G. Let $D^* = D \cap R$. For $d \in D$, $\mathrm{ad}(d)$ takes the complete lattice D^* onto itself. Thus the subspace $L(d) = (\mathrm{ad}(d) - I)R$ has a basis consisting of elements of D^*. Therefore, for each $d \in D$, $L(d)/L(d) \cap D$ is compact. Hence the linear space spanned by all the subspaces $L(d)$, i.e., $[D,R]$, contains sufficiently many elements of D^* to make $[D,R]/D \cap [D,R]$ compact.

Now $[D,R]$ is a linear subspace of R which clearly equals $[\exp \mathfrak{G}^*,R]$, where $\mathfrak{G}^*$ is the subset of $\mathfrak{G}$ defined in the lemma. But the conclusion of that lemma, together with the fact that linear subspaces in a finite dimensional space are automatically closed, tells us that $[\exp \mathfrak{G}^*,R] = [G,R]$. This completes the proof of the theorem.

COROLLARY. $G^\infty/G^\infty \cap D$ is compact.

PROOF. Without loss of generality, we may assume that G is not nilpotent, so G^∞ is non-trivial. Let N equal the maximal normal analytic subgroup of G. Since $N \supset [G,G]$, it is easily seen that

$$G^\infty = [G,\ldots[G,[G,N]]\ldots] .$$

But now we may apply the theorem inductively, since $N \cap D$ is uniform in N.

CHAPTER VIII

AN APPLICATION OF NILFLOWS TO

DIOPHANTINE APPROXIMATIONS

by

L. Auslander and F. Hahn

In this chapter we will use the results established in Chapters IV and V to obtain a generalization of the Kronecker approximation theorem (Koksma [1]). We point out that the theorem proved here is similar to theorem 14 in Weyl [1]. The major difference is in the fact that the set P is relatively dense. In order to facilitate the exposition we use the following notational convention: If x is a real number we say that $|x| < \varepsilon$, mod. 1, if there is an integer q such that $|x-q| < \varepsilon$. We may now state Theorem 1 as follows:

THEOREM 1. Let $p_i(x) = \sum_{j=1}^{n} a_{ij} x^j$, where the a_{ij} are integers, and $\sum_j |a_{ij}| > 0$ for each $i = 1,2\ldots,n$. Let λ_i be a set of n real numbers such that $1,\lambda_1,\ldots,\lambda_n$ are rationally independent. If $\varepsilon > 0$ is given and if θ_i, $i = 1,2,\ldots,n$ are arbitrary real numbers, then there is a relatively dense set P of the integers such that

$$|\lambda_i p_i(c) - \theta_i| < \varepsilon, \quad \text{mod } 1 \quad ,$$

for $i = 1,2,\ldots,n$ and $c \in P$.

To prove this theorem we study a particular type of nilmanifold and a discrete flow on it. The discrete flow comes from restricting a one-parameter flow to integral values of the parameter. A generic group of our type will be a matrix group N with elements of the form

$$\begin{pmatrix} N_1 & & & & 0 \\ & N_2 & & & \\ & & \cdot & & \\ & & & \cdot & \\ 0 & & & & \cdot \\ & & & & N_n \end{pmatrix}$$

where each N_i is of the form

$$\begin{pmatrix} 1 & x & x^2/2! & & & x^{n-1}/(n-1)! & y_{ni} \\ & 1 & x & & & x^{n-2}/(n-2)! & y_{n-1\,i} \\ & & 1 & \cdot & & \vdots & \vdots \\ & & & \cdot & & \vdots & \vdots \\ & & & & 1 & x & y_{2i} \\ & 0 & & & & 1 & y_{1i} \\ & & & & & & 1 \end{pmatrix}$$

A typical element of N will be written (x,y_{ji}). Let D be the discrete subgroup of N whose elements are of the form $(n!a,b_{ji})$, where a,b_{ji}, $i,j = 1,\ldots,n$, are arbitrary integers. It follows from Malcev [1] that $D\backslash N$ is compact.

In order to study any flows on $D\backslash N$ we must be familiar with the matrix multiplication of N. It has the following form

$$(x,y_{ij})(z,w_{ji}) = (v,u_{ji})$$

where

$$v = x + z$$
$$u_{1i} = y_{1i} + w_{1i}$$
$$u_{2i} = y_{2i} + w_{2i} + w_{1i}x$$
$$u_{ji} = y_{ji} + w_{ji} + w_{j-1,i}x + w_{j-2i}x^2/2! + \ldots + w_{1i}x^{j-1}/(j-1)!$$

for $i,j = 1,2,\ldots,n$.

By a straightforward computation we see that

$$\varphi(t) = (t,\lambda_i t^j/j!) \quad i,j = 1,2,\ldots,n \ .$$

is a one-parameter subgroup in N, where t is the group parameter and the λ_i are constants. When necessary we will also denote this one-parameter subgroup by $\varphi(t;\lambda_i)$.

The idea of the proof of theorem 1 is not difficult but quite computational in nature. We observe first that under the hypotheses of the theorem the flow $(D\backslash N,\varphi(t;\lambda_i))$ is minimal. We then examine the orbit of the coset D for integral values of t. These observations are interpreted in terms of the coordinates of the group N. This interpretation forms the proof of the theorem. For the sake of clarity our line of reasoning is divided into two lemmas. In both these lemmas we make use of the notation of the previous discussion.

LEMMA 1. In N let $d = (n!a,b_{ji})$ be an element of D and let $c = -n!a$. If $(v,u_{ji}) = d\cdot\varphi(c;\lambda_i)$ then

$$v = n!a + c$$
$$u_{ji} = b_{ji} + (-1)^{j+1}\lambda_i c^j/j! \ .$$

PROOF. We use the multiplication table for N and the binomial expansion of $0 = (1-1)^j$ to obtain our result

$$\begin{aligned}
v &= n!a + c \\
u_{ji} &= b_{ji} + \lambda_i c^j/j! + \dots + \lambda_i c^{j-k}(n!a)^k/k!(j-k)! \\
&\qquad + \dots + \lambda_i c(n!a)^{j-1}/(j-1)! \\
&= b_{ji} + \lambda_i c^j \sum_{k=0}^{j-1} (-1)^k/(j-k)!\ k! \\
&= b_{ji} + \frac{1}{j!}(\lambda_i c^j \sum_{k=0}^{j} (-1)^k j!/(j-k)!\ k!) + \\
&\qquad + (-\lambda_i c^j(-1)^j/j!) \\
&= b_{ji} + (-1)^{j+1}\lambda_i c^j/j! \quad .
\end{aligned}$$

We will now adopt the following notation. If $\alpha > 0$ is given and if β_{ji}, $i,j = 1,2,\dots,n$ is a set of any n^2 real numbers then $V(\alpha) = V(\alpha,\beta_{ji})$ is the subset of N defined by

$$V(\alpha,\beta_{ji}) = \{(x,y_{ji}) : |x| < \alpha,\ |y_{ji} - \beta_{ji}| < \alpha,\ i,j = 1,2,\dots,n\} .$$

LEMMA 2. Let $\alpha > 0$ be given and let β_{ji}, $i,j = 1,2,\dots,n$, be a set of n^2 arbitrary real numbers. Let λ_i be n real numbers such that $1,\lambda_1,\dots,\lambda_n$ are rationally independent. Under these conditions there is a relatively dense set P of integers which has the property that $c \in P$ implies there is a $d \in D$ for which $d\varphi(c;\lambda_i) \in V(\alpha)$.

PROOF. There is a δ, $0 < \delta < \min(\alpha/2, \frac{1}{2})$ such that $V(\alpha/2)\varphi(t;\lambda_i) \subset V(\alpha)$, whenever $|t| < \delta$. Corollary 4.5 of Chapter V asserts that $(D\backslash N, \varphi(t;\lambda_i))$ is a minimal set and thus each orbit is almost periodic (Gottschalk-Hedlund [1]). Thus there is a relatively dense subset S of the reals having the property that $s \in S$ implies there is a $d \in D$ for which $d\varphi(s;\lambda_i) \in V(\delta)$. Since $d = (n!a, b_{ji})$ we see that $|n!a - s| < \delta$. Consequently there is a real number t such that $t + s = n!a$ and $|t| < \delta$. This implies that $d\varphi(s + t;\lambda_i) \in V(\delta)\varphi(t;\lambda_i) \subset V(\alpha)$. Let P be the set of all integers obtained from S in the preceding fashion. Since S is relatively dense so is P and P satisfies the conclusion of the theorem.

PROOF of THEOREM 1. In Lemmas 1 and 2 let

$\alpha = \min(\varepsilon/(n \cdot n! \cdot \sum_{ji} |a_{ji}|), 1)$ and let n_i be the number of non-zero a_{ji} for each i, and let

$$\beta_{ji} = \theta_i/n_i(-1)^{j+1} a_{ji}\ j! \quad \text{if} \quad a_{ji} \neq 0$$

$$\beta_{ji} = 0 \quad \text{if} \quad a_{ji} = 0 \ .$$

According to Lemma 2 there is a relatively dense set of integers P having the property $c \in P$ implies there is a $d \in D$ such that $d\varphi(c;\lambda_i) \in V(\alpha)$.

Thus if $d = (n!a, b_{ji})$ and $d\varphi(c;\lambda_i) = (v, u_{ji})$ we see that $|v| = |n!a + c| < \alpha$. But $\alpha < 1$, $n!a$, and c are integers, so $c = -n!a$. Applying Lemma 1 we obtain

$$u_{ji} = b_{ji} + (-1)^{j+1} \lambda_i c^j/j! \ .$$

Since $|u_{ji} - \beta_{ji}| < \alpha$, we see

$$|b_{ji}(-1)^{j+1} j! + \lambda_i c^j - (-1)^{j+1} j! \beta_{ji}| < \alpha j! \leq \alpha n! \ .$$

Multiplying this by $|a_{ji}|$ and summing over those j for which $a_{ji} \neq 0$, we obtain

$$|\sum_j a_{ji} b_{ji}(-1)^{j+1} j! + \lambda_i p(c) - \theta_i| \leq$$

$$\sum_j |a_{ji}| |b_{ji}(-1)^{j+1} j! + \lambda_i c^j - (-1)^{j+1} j! \beta_{ji}|$$

$$< n\alpha n! \sum_j |a_{ji}| < \varepsilon \ .$$

Since a_{ji} and b_{ji} are integers we have

$$|\lambda_i p(c) - \theta_i| < \varepsilon, \ \text{mod. } 1 \quad \text{for } i = 1,\ldots,n \text{ and } c \in P \ .$$

The next theorem is a special case of a theorem of van der Corput [1]. The reader should be warned that it is quite a different theorem than Theorem 1. The hypotheses of this theorem will assert nothing about the rational independence of the λ_i. The conclusion in turn will speak only about approximations to zero, mod. 1.

THEOREM 2. Let $p_i(x) = \sum_{j=1}^{n} a_{ji} x^j$, a_{ji} integers, and $\sum_{j=1}^{n} |a_{ji}| > 0$ for $i = 1,2,\ldots,n$. Let $\lambda_1,\ldots,\lambda_n$ be a set of any n real numbers. If $\varepsilon > 0$ is given then there is a relatively dense set of integers P such that

$$|\lambda_i p_i(c)| < \varepsilon, \ \text{mod. } 1$$

for $c \in P$ and $i = 1,2,\ldots,n$.

PROOF. We observe that Lemma 2 is true if β_{ji} are all zero and the restriction of rational independence of 1 and the λ_i is removed. To see this we use the fact that $(D\backslash N, \varphi(t;\lambda_i))$ is always pointwise periodic (Theorem 4, Chapter IV). We now merely restrict our attention to the orbit of the coset D and the proof of Lemma 2 goes through in the same fashion. Lemma 1 remains unaltered and the proof of Theorem 2 is now straight-forward.

CHAPTER IX.

DISCRETE GROUPS WITH DENSE ORBITS

by

Leon Greenberg

§1. Introduction. Let $\mathbf{D}$ be a division ring (which will immediately be specialized to $\mathbf{R}$ = the real field, $\mathbf{C}$ = the complex field or $\mathbf{Q}$ = the quaternions) and let $\mathbf{D}_n$ denote the ring of $n \times n$ matrices with elements in $\mathbf{D}$. The general linear group $GL(n,\mathbf{D})$ is the group of non-singular matrices in $\mathbf{D}_n$; the special linear group $SL(n,\mathbf{D})$ is the group of matrices with determinant 1 (in the sense of J. Dieudonne [1]); if $\mathbf{D}$ is a field, the simplectic group $Sp(2n,\mathbf{D})$ is the group of matrices which leaves invariant the bilinear form $(x,y) = \sum_{i=1}^{n} (x_i y_{i+n} - x_{i+n} y_i)$.

We shall prove (a more general version of) the following theorem.

THEOREM 1. Let $\mathbf{D} = \mathbf{R}$, $\mathbf{C}$ or $\mathbf{Q}$, let V be an n-dimensional vector space over $\mathbf{D}$ and let G be a (real) Lie subgroup of $GL(n,\mathbf{D})$. Suppose that either

a) $G \supset SL(n,\mathbf{D})$, $n > 1$, or

b) $G \supset Sp(n,\mathbf{D})$, n even, $\mathbf{D} = \mathbf{R}$ or $\mathbf{C}$.

Let Γ be a discrete subgroup of G with compact factor space $\Gamma\backslash G$. Then if $v \in V$, $v \neq 0$, the orbit Γv is dense in V.

The above theorem will actually be proved for a certain class of groups G which have sufficient transitivity properties in V.

Note that if $H = \{g \in G | g(v) = v\}$, Γv dense in V is equivalent to ΓH dense in G. The theorem can be reinterpreted as follows: Let M be any compact manifold with a G-structure (i.e., M has a homogeneous covering space G/K, where K is compact, and the group of covering transformations is a discrete subgroup Γ of G). Then H is immersed in M (under the map $H \subset G \longrightarrow G/K \longrightarrow M$) as an everywhere dense subset. This is an analogue of a theorem of G. A. Hedlund [1] that every curve of geodesic curvature 1 in a compact surface S with constant negative curvature is dense in S. The present theorem coincides with Hedlund's theorem for the case $G = SL(2,\mathbf{R})$.

We might also remark that K. Mahler [1] has proved the following theorem.

Let

$$\Gamma = \left\{\gamma^{(k)} \,\middle|\, \gamma^{(k)}(z) = \frac{\gamma_{11}^{(k)}\; z + \gamma_{12}^{(k)}}{\gamma_{21}^{(k)}\; z + \gamma_{22}^{(k)}},\ \gamma_{ij}^{(k)} \text{ real},\ \gamma_{11}^{(k)}\gamma_{22}^{(k)} - \gamma_{11}^{(k)}\gamma_{21}^{(k)} = 1\right\}$$

be a Fuchsian group with compact fundamental region. Let $f(x,y)$ *be a positive definite form. Then the values* $f(\gamma_{11}^{(k)}, \gamma_{21}^{(k)})$ *are everywhere dense in the positive real numbers.*

Mahler also conjectured that if f is an indefinite form, the values $f(\gamma_{11}^{(k)}, \gamma_{21}^{(k)})$ are dense in the real numbers; if Γ is a Kleinian polyhedral group and f a (positive definite) Hermitian form, then the values $f(\gamma_{11}^{(k)}, \gamma_{21}^{(k)})$ are dense in the (positive) real numbers. Theorem 1 not only verifies these conjectures but shows that if $\Gamma = \{\gamma^{(k)} = (\gamma_{ij}^{(k)})\}$ satisfies the hypotheses of Theorem 1, and $f(x_1,\ldots,x_n)$ is any form, then the values $f(\gamma_{11}^{(k)}, \gamma_{21}^{(k)},\ldots,\gamma_{n1}^{(k)})$ are dense in the range of f. For if v is the vector $(1,0,\ldots,0)$, then $\gamma^{(k)}(v) = (\gamma_{11}^{(k)}, \gamma_{21}^{(k)},\ldots,\gamma_{n1}^{(k)})$. Theorem 1 says that the n-tuples $(\gamma_{11}^{(k)}, \gamma_{21}^{(k)},\ldots,\gamma_{n1}^{(k)})$ are dense in V. Therefore, if f is any continuous function on V, the values $f(\gamma_{11}^{(k)},\ldots,\gamma_{n1}^{(k)})$ are dense in the range of f.

§2. *Strongly Transitive Groups*. In this section $\mathbf{D}$ is $\mathbf{R}$ = the real field, $\mathbf{C}$ = the complex field or $\mathbf{Q}$ = the quaternions, and V is an n-dimensional right vector space over $\mathbf{D}$.

If $A \in \mathbf{D}_n$, an eigenvalue λ of A will mean (as usual) an element $\lambda \in \mathbf{D}$ ($\lambda \in \mathbf{C}$ if $\mathbf{D} = \mathbf{R}$) for which there is a vector $v \neq 0$ (in the complexification of V if $\mathbf{D} = \mathbf{R}$) such that $Av = v\lambda$. In the case $\mathbf{D} = \mathbf{Q}$, A can have a large set of eigenvalues, for if λ is an eigenvalue, then $d^{-1}\lambda d$ is also. However, it may be verified (by considering V as a 2n-dimensional space over $\mathbf{C}$) that A has a set of n eigenvalues $\lambda_1,\ldots,\lambda_n$ (each listed with possible multiplicities) such that any eigenvalue of A is conjugate to some λ_i. We shall call $(\lambda_1,\ldots,\lambda_n)$ a *representative set* of eigenvalues for A.

A will be called semi-simple if it has a representative set of eigenvalues $(\lambda_1,\ldots,\lambda_n)$ for which there is a basis $v_1,\ldots,v_n$ of V with $Av_\alpha = v_\alpha\lambda_\alpha$ $(\alpha = 1,\ldots,n)$. This will be true, in particular, if no pair $\lambda_\alpha,\lambda_\beta$ are conjugates. The basis $(v_1,\ldots,v_n)$ will be called an *eigenbasis* for A.

If $v \neq 0$, the *ray through* v is the set $\rho(v) = \{w \mid w = vr,$ r real and $> 0\}$. V has a natural topology (as a vector space over R) and this induces a quotient topology on the space of rays ($\rho(v)$ is an equivalence class under the relation: $w \sim v$ if $w = vr$, r real and > 0).

DEFINITION. A subgroup $G \subset GL(n,\mathbf{D})$ is called *strongly transitive* if G satisfies the following conditions.

ST 1. G is transitive: If v and w are non-zero vectors, then there exists $g \in G$ such that $g(v) = w$.

ST 2. G is slmost doubly transitive on rays:

a) If ρ_1, ρ_2 are two rays and Ω_1, Ω_2 are two open sets of rays, then there is an element $g \in G$ such that $g\rho_1 \in \Omega_1$ and $g^{-1}\rho_2 \in \Omega_2$.

b) If ρ_1, ρ_2 are two rays and Ω is an open set of rays, there exists an element $g \in G$ such that $g\rho_1$ and $g\rho_2 \in \Omega$.

ST 3. G contains an element g which has a representative set of eigenvalues $(\lambda_1, \ldots, \lambda_n)$ such that:

a) The norms $|\lambda_1|, \ldots, |\lambda_n|$ are distinct.

b) $|\lambda_1| > 1 > |\lambda_2|$.

Note that $SL(n, \mathbf{D})$ and $Sp(2n, \mathbf{D})$ ($n > 1$ in the first case, $\mathbf{D} = \mathbf{R}$ or $\mathbf{C}$ in the second case) are strongly transitive groups. Moreover, strong transitivity is inherited by larger groups in $GL(n, \mathbf{D})$.

The theorem we wish to prove is the following.

THEOREM 2. Let G be a locally compact, strongly transitive group, and let Γ be a discrete subgroup of G with compact factor space $\Gamma\backslash G$. Then for any non-zero vector $v \in V$, the orbit Γv is dense in V.

The proof will be broken up into several lemmas. We shall suppose that a fixed Euclidean norm is chosen in V with respect to some basis $v_1, \ldots, v_n$: If $v = \sum_{\alpha=1}^{n} v_\alpha d_\alpha$, then $|v| = \left(\sum_{\alpha=1}^{n} |d_\alpha|^2\right)^{\frac{1}{2}}$. If X is any subset of V,

$$|v - X| = \inf_{x \in X} |v - x| .$$

LEMMA 1. Let $A \in \mathbf{D}_n$ and let $(\lambda_1, \ldots, \lambda_n)$ be a representative set of eigenvalues for A. Suppose that $|\lambda_1| \neq |\lambda_\alpha|$ $(\alpha = 2, \ldots, n)$ and v_1 is an eigenvector corresponding to the eigenvalue λ_1. For $\varepsilon > 0$ there is a neighborhood Ω of A such that if $B \in \Omega$, then B has a representative set of eigenvalues $(\mu_1, \ldots, \mu_n)$ and an eigenvector w_1 corresponding to the eigenvalue μ_1 with

1) $|\mu_\alpha - \lambda_\alpha| < \varepsilon$ $(\alpha = 1, \ldots, n)$

2) $|w_1 - v_1| < \varepsilon$.

PROOF. This is known for the cases $\mathbf{D} = \mathbf{R}$ or $\mathbf{C}$. For $\mathbf{D} = \mathbf{Q}$, it can be proved by considering V as a $2n$-dimensional vector space over $\mathbf{C}$.

If $A = (a_{\alpha\beta}) \in \mathbf{D}_n$, we consider A as a linear transformation on V by choosing some fixed basis $t_1, \ldots, t_n$ of V and letting $At_\beta = \sum_{\alpha=1}^{n} t_\alpha a_{\alpha\beta}$. If $t = \sum_{\alpha=1}^{n} t_\alpha d_\alpha$, then $At = \sum_{1 \le \alpha, \beta \le n} t_\alpha a_{\alpha\beta} d_\beta$.

Define $\|A\| = \sup_{|t|=1} |At|$. Then $|At| \leq \|A\| \, |t|$,

$\|A + B\| \leq \|A\| + \|B\|$ and $\|AB\| \leq \|A\| \, \|B\|$. If $\mathbf{S} \subset \mathbf{D}_n$, define $\|A - \mathbf{S}\| = \inf_{B \in \mathbf{S}} \|A - B\|$.

LEMMA 2. Let $\mathbf{S}$ be the set of all matrices $A(\xi) = \operatorname{diag}(\xi,0,0,\ldots,0)$ with $|\xi| = 1$. The matrices $A(\xi)$ all have a common eigenbasis $(v_1,\ldots,v_n)$.

a) For $\varepsilon > 0$ there is an open set Ω containing $\mathbf{S}$ such that if $B = (b_{\alpha\beta}) \in \Omega$ then, B has a representative set of eigenvalues $(\mu_1,\ldots,\mu_n)$ and an eigenvector w_1 corresponding to μ_1 with

$$1) \quad |\mu_1 - b_{11}| < \varepsilon ,$$
$$2) \quad |\mu_\alpha| < \varepsilon \quad (\alpha = 2,\ldots,n) ,$$
$$3) \quad |w_1 - v_1| < \varepsilon$$

b) Any open set Ω which contains $\mathbf{S}$ contains an open set of the form $\Sigma(\mathbf{S},\delta) = \{B \mid \|B - \mathbf{S}\| < \delta\}$.

c) If Ω is an open set containing $\mathbf{S}$, then there is an open set $\Omega_{\mathbf{S}}$ containing $\mathbf{S}$ and an open set Ω_1 containing 1 so that $\Omega_1\Omega_{\mathbf{S}} \subset \Omega$.

PROOF. a) By Lemma 1, for each ξ there is an open set $\Omega(\xi)$ containing $A(\xi)$ so that $B \in \Omega(\xi)$ has a representative set of eigenvalues $(\mu_1,\ldots,\mu_n)$ and an eigenvector w_1 corresponding to μ_1 with

$$4) \quad |\mu_1 - \xi| < \frac{\varepsilon}{2} ,$$
$$5) \quad |\mu_\alpha| < \varepsilon , \qquad \alpha = 2,\ldots,n ,$$
$$6) \quad |w_1 - v_1| < \varepsilon .$$

Let $\Omega(\xi)$ also fulfill the condition

$$7) \quad |b_{11} - \xi| < \frac{\varepsilon}{2} .$$

Then $\Omega = \bigcup_{\xi \in \mathbf{S}} \Omega(\xi)$ has the required property.

b) For each ξ, there is $\delta(\xi)$ so that $\Sigma(\xi) = \{B \mid \|B-A(\xi)\| < \delta(\xi)\}$ is contained in Ω. $\mathbf{S}$ is covered by the sets $\Sigma'(\xi) = \{B \mid \|B-A(\xi)\| < \frac{1}{2}\delta(\xi)\}$. Since $\mathbf{S}$ is compact, $\mathbf{S}$ is covered by a finite number of these neighborhoods: $\Sigma'(\xi_1),\ldots,\Sigma'(\xi_n)$. Let $\delta = \frac{1}{2} \min_{1 \leq \alpha \leq m} \delta(\xi_\alpha)$. Then for this value of δ, $\Sigma(\mathbf{S},\delta) \subset \Omega$. For if $B \in \Sigma(\mathbf{S},\delta)$ then $\|B-\mathbf{S}\| < \delta$. Since $\mathbf{S}$ is compact, there is $A(\xi)$ so that $\|B-\mathbf{S}\| = \|B-A(\xi)\|$. Moreover, there is ξ_α so that $A(\xi) \in \Sigma'(\xi_\alpha)$, or $\|A(\xi) - A(\xi_\alpha)\| < \frac{1}{2}\delta(\xi_\alpha)$. Then $\|B-A(\xi_\alpha)\| \leq \|B-A(\xi)\| + \|A(\xi) - A(\xi_\alpha)\| < \delta + \frac{1}{2}\delta(\xi_\alpha) \leq \frac{1}{2}\delta(\xi_\alpha) + \frac{1}{2}\delta(\xi_\alpha) = \delta(\xi_\alpha)$. Thus $B \in \Sigma(\xi_\alpha) \subset \Omega$.

c) Since the map $(A,B) \longrightarrow AB$ is continuous, for each $\xi(|\xi| = 1)$ there is a neighborhood $\Omega(\xi)$ of $A(\xi)$ and a neighborhood

$\Omega_\xi(1)$ of 1 so that $\Omega_\xi(1)\Omega(\xi) \subset \Omega$. Since $\mathbf{S}$ is compact, $\mathbf{S}$ is covered by a finite number of these neighborhoods: $\Omega(\xi_1),\ldots,\Omega(\xi_m)$. Let $\Omega_{\mathbf{S}} = \bigcup_{1 \leq \alpha \leq m} \Omega(\xi_\alpha)$, $\Omega_1 = \bigcap_{1 \leq \alpha \leq m} \Omega_{\xi_\alpha}(1)$. Then

$$\Omega_1\Omega_{\mathbf{S}} \subset \bigcup_{1 \leq \alpha \leq m} \Omega_{\xi_\alpha}(1)\, \Omega(\xi_\alpha) \subset \Omega \quad .$$

LEMMA 3. Let A be a semi-simple element in $GL(n,\mathbf{D})$, which has a representative set of eigenvalues $(\lambda_1,\ldots,\lambda_n)$ and a corresponding eigenbasis $(v_1,\ldots,v_n)$ such that $|\lambda_1|,\ldots,|\lambda_n|$ are distinct. For $\varepsilon > 0$ there is a neighborhood Ω of 1 in $GL(n,\mathbf{D})$ such that (for all integers $r > 0$) if $B \in \Omega A^r$, then B is semi-simple and has a representative set of eigenvalues $(\mu_1,\ldots,\mu_r)$ and a corresponding eigenbasis $(w_1,\ldots,w_n)$ with

1) $|\mu_\alpha - \lambda_\alpha^r| < \varepsilon|\lambda_\alpha|^r \quad (\alpha = 1,\ldots,n)$,

2) $|w_\alpha - v_\alpha| < \varepsilon \quad , \quad (\alpha = 1,\ldots,n)$.

PROOF. We shall use induction on n. For $n = 1$, the statement is easily verified. We may suppose that $|\lambda_1| > |\lambda_2| > \cdots > |\lambda_n|$, and (by using the basis $v_1,\ldots,v_n$ to represent linear transformations by matrices) that $A = \operatorname{diag}(\lambda_1,\ldots,\lambda_n)$.

$$\text{Let } A_1 = \frac{1}{|\lambda_1|} A = \operatorname{diag}\left(\frac{\lambda_1}{|\lambda_1|}, \frac{\lambda_2}{|\lambda_1|}, \ldots, \frac{\lambda_n}{|\lambda_1|}\right) .$$

$\|A_1^r - \mathbf{S}\| \longrightarrow 0$ as $r \longrightarrow \infty$ ($\mathbf{S}$ as in Lemma 2).

Choose a number $\delta > 0$. By Lemma 2a there is an open set Ψ containing $\mathbf{S}$ so that if $B_1 = (b_{\alpha\beta}) \in \Psi$, then B_1 has an eigenvalue μ_1' and a corresponding eigenvector w_1 with

3) $|\mu_1' - b_{11}| < \frac{\delta}{2}$

4) $|w_1 - v_1| < \delta$.

By Lemma 2c) there is an open set $\Omega_{\mathbf{S}}$ containing $\mathbf{S}$ and an open set Ω_1 containing 1 so that $\Omega_1\Omega_{\mathbf{S}} \subset \Psi$. Since $\|A_1^r - \mathbf{S}\| \longrightarrow 0$ as $r \longrightarrow \infty$, Lemma 2b tells us that there is an integer r_0 so that $A_1^r \in \Omega_{\mathbf{S}}$ for $r > r_0$.

Now suppose that $B \in \Omega_1 A^r$ $(r > r_0)$. $B = \omega A^r$, where $\omega = (\varepsilon_{\alpha\beta}) \in \Omega_1$. Let $B_1 = \frac{1}{|\lambda_1|^r} B$. Then $B_1 = \omega A_1^r \in \Omega_1 A_1^r \subset \Omega_1\Omega_{\mathbf{S}} \subset \Psi$. Therefore B_1 has an eigenvalue μ_1' and a corresponding eigenvector w_1 so that 3) and 4) are satisfied (where $b_{11} = \varepsilon_{11} \dfrac{\lambda_1^r}{|\lambda_1|^r}$). Then B has an eigenvalue $\mu_1 = |\lambda_1|^r \mu_1'$ with corresponding eigenvector w_1, and

$$|\mu_1 - \lambda_1^r| \le |\mu_1 - |\lambda_1|^r b_{11}| + | |\lambda_1|^r b_{11} - \lambda_1^r|$$

$$= | |\lambda_1|^r \mu_1' - |\lambda_1|^r b_{11}| + |\varepsilon_{11}\lambda_1^r - \lambda_1^r|$$

$$< \frac{\delta}{2} |\lambda_1|^r + |\varepsilon_{11} - 1| |\lambda_1|^r .$$

If we choose Ω_1 small enough so that $|\varepsilon_{11} - 1| < \frac{\delta}{2}$, then $|\mu_1 - \lambda_1^r| < \delta|\lambda_1|^r$. We now know that if $r > r_0$ and $B \in A^r\Omega_1$ then B has an eigenvalue μ_1 and corresponding eigenvector w_1 such that

5) $|\mu_1 - \lambda_1^r| < \delta|\lambda_1|^r$

6) $|w_1 - v_1| < \delta$.

By Lemma 1, Ω_1 can be shosen small enough so that this is also true for $r = 1,2,\ldots,r_0$.

Now consider the $(n-1) \times (n-1)$ matrix $A_2 = \mathrm{diag}(\lambda_2,\lambda_3,\ldots,\lambda_n)$. By the inductive hypothesis, there is a neighborhood $\Omega^{(n-1)}$ of 1 in $\mathbf{D}_{n-1}$ so that (for all integers $r > 0$) if $B \in \Omega^{(n-1)}A_2^r$, then B has eigenvalues $\mu_2,\ldots,\mu_n$ and corresponding eigenvectors $w_2',\ldots,w_n'$ with

7) $|\mu_\alpha - \lambda_\alpha^r| < \delta|\lambda_\alpha|^r \qquad (\alpha = 2,\ldots,n)$,

8) $|w_\alpha' - v_\alpha| < \delta \qquad (\alpha = 2,\ldots,n)$.

Let Ω_2 be a compact neighborhood of 1 in $\mathbf{D}_n$ such that $(\varepsilon_{\alpha\beta}) \in \Omega_2$ implies that

$$\begin{pmatrix} \varepsilon_{22} & \cdots & \varepsilon_{2n} \\ \cdot & & \\ \cdot & & \\ \cdot & & \\ \varepsilon_{n2} & \cdots & \varepsilon_{nn} \end{pmatrix} \in \Omega^{(n-1)} .$$

Let $X = \{x \in GL(n,\mathbf{D}) \mid |x(v_1) - v_1| \le \delta,\ x(v_\alpha) = v_\alpha,\ \alpha = 2,\ldots,n\}$. We assert that there is a neighborhood Ω of 1 in $GL(n,\mathbf{D})$ such that

9) $x^{-1}\Omega\, x \subset \Omega_2$ for all $x \in X$,

10) $\Omega \subset \Omega_1$.

The map $(x,y) \longrightarrow x^{-1}y\,x$ $(x,y \in GL(n,D))$ is continuous. Therefore for each $x \in X$ there is a neighborhood $\Omega(x)$ of x and a neighborhood $\Omega_x(1)$ of 1 so that $m^{-1}nm \in \Omega_2$ for $m \in \Omega(x)$, $n \in \Omega_x(1)$. Since X is compact, X is covered by a finite number of these neighborhoods: $\Omega(x_1),\ldots,\Omega(x_r)$. Let $\Omega = \Omega_1 \cap \bigcap_{1 \le \alpha \le r} \Omega_{x_\alpha}(1)$. Then Ω satisfies 9 and 10

Now let $\omega \in \Omega$ and $B = \omega A^r$. We know that B has an eigenvalue μ_1 with corresponding eigenvector w_1 so that

5) $|\mu_1 - \lambda_1^r| < \delta |\lambda_1|^r$

6) $|w_1 - v_1| < \delta$.

There is $x \in X$ so that $x(v_1) = w_1$. $x^{-1}Bx$ has eigenvalue μ_1 with corresponding eigenvector $x^{-1}(w_1) = v_1$. $x(v_1) = \sum_{\alpha=1}^{n} v_\alpha x_\alpha$ and $x(v_\alpha) = v_\alpha$, $\alpha > 1$. Thus x is of the form:

$$x = \begin{pmatrix} x_1 & & & & \\ x_2 & 1 & & & \\ x_3 & & 1 & & \\ \cdot & & & \cdot & \\ \cdot & & & & \cdot \\ \cdot & & & & & \cdot \\ x_n & & & & & & 1 \end{pmatrix} .$$

Since $x^{-1}A^r x\,(v_\alpha) = v_\alpha \lambda_\alpha^r$ (for $\alpha > 1$),

$$x^{-1}A^r x = \begin{pmatrix} a_{11} & & & & \\ a_{21} & \lambda_2^r & & & \\ a_{31} & & \lambda_3^r & & \\ \cdot & & & \cdot & \\ \cdot & & & & \cdot \\ \cdot & & & & & \cdot \\ a_{n1} & & & & & & \lambda_n^r \end{pmatrix}$$

$x^{-1}\omega x = (\varepsilon_{\alpha\beta}) \in \Omega_2$, therefore

$$\begin{pmatrix} \varepsilon_{22} & \cdots & \varepsilon_{2n} \\ \cdot & & \\ \cdot & & \\ \cdot & & \\ \varepsilon_{n2} & \cdots & \varepsilon_{nn} \end{pmatrix} \in \Omega^{(n-1)} .$$

$$x^{-1}Bx = (x^{-1}\omega x)(x^{-1}A^r x) = \begin{pmatrix} \varepsilon_{11} & \cdots & \varepsilon_{1n} \\ \cdot & & \\ \cdot & & \\ \cdot & & \\ \varepsilon_{n1} & \cdots & \varepsilon_{nn} \end{pmatrix} \begin{pmatrix} a_{11} & & & & \\ a_{21} & \lambda_2^r & & & \\ \cdot & & \cdot & & \\ \cdot & & & \cdot & \\ \cdot & & & & \cdot \\ a_{n1} & & & & & \lambda_n^r \end{pmatrix}$$

$$= \begin{pmatrix} b_{11} & \varepsilon_{12}\lambda_2^r & \cdots & \varepsilon_{1n}\lambda_n^r \\ b_{21} & \varepsilon_{22}\lambda_2^r & \cdots & \varepsilon_{2n}\lambda_n^r \\ \cdot & \cdot & & \\ \cdot & \cdot & & \\ \cdot & \cdot & & \\ b_{n1} & \varepsilon_{1n}\lambda_2^r & \cdots & \varepsilon_{nn}\lambda_n^r \end{pmatrix} .$$

But since $x^{-1}Bx$ has eigenvalue μ_1 with corresponding eigenvector v_1, we have $b_{11} = \mu_1$, $b_{\alpha 1} = 0$ for $\alpha > 1$, and

$$x^{-1}Bx = \begin{pmatrix} \mu_1 & \varepsilon_{12}\lambda_2^r & \cdots & \varepsilon_{1n}\lambda_n^r \\ 0 & \varepsilon_{22}\lambda_2^r & \cdots & \varepsilon_{2n}\lambda_n^r \\ \cdot & & & \\ \cdot & & & \\ \cdot & & & \\ 0 & \varepsilon_{n2}\lambda_2^r & \cdots & \varepsilon_{nn}\lambda_n^r \end{pmatrix}$$

Since $\begin{pmatrix} \varepsilon_{22} & \cdots & \varepsilon_{2n} \\ \cdot & & \\ \cdot & & \\ \cdot & & \\ \varepsilon_{n2} & \cdots & \varepsilon_{nn} \end{pmatrix} \in \Omega^{(n-1)}$ the matrix

$$\begin{pmatrix} \varepsilon_{22} & \cdots & \varepsilon_{2n} \\ \cdot & & \cdot \\ \cdot & & \cdot \\ \cdot & & \cdot \\ \varepsilon_{n2} & \cdots & \varepsilon_{nn} \end{pmatrix} A_2^r = \begin{pmatrix} \varepsilon_{22}\lambda_2^r & \cdots & \varepsilon_{2n}\lambda_n^r \\ \cdot & & \\ \cdot & & \\ \cdot & & \\ \varepsilon_{n2}\lambda_2^r & \cdots & \varepsilon_{nn}\lambda_n^r \end{pmatrix}$$

has eigenvalues $\mu_2,\ldots,\mu_n$ with corresponding eigenvectors $w_2',\ldots,w_n'$ satisfying

7) $|\mu_\alpha - \lambda_\alpha^r| < \delta|\lambda_\alpha|^r \qquad (\alpha = 2,\ldots,n)$

8) $|w_\alpha' - v_\alpha| < \delta \qquad (\alpha = 2,\ldots,n)$.

Let $w_\gamma' = \sum_{\alpha=2}^{n} v_\alpha y_{\alpha\gamma}$, $\gamma = 2,\ldots,n$. Then $\sum_{\beta=2}^{n} \varepsilon_{\alpha\beta}\lambda_\beta^r y_{\beta\gamma} = y_{\alpha\gamma}\mu_\gamma$ $(\alpha,\gamma = 2,\ldots,n)$. The vector $w_\gamma'' = v_1 y_{1\gamma} + w_\gamma' = \sum_{\alpha=1}^{n} v_\alpha y_{\alpha\gamma}$ will be an eigenvector of $x^{-1}Bx$ with corresponding eigenvalue μ_γ, provided that $y_{1\gamma}$ satisfies the equation

11) $\mu_1 y_{1\gamma} + \sum_{\beta=2}^{n} \varepsilon_{1\beta}\lambda_\beta^r y_{\beta\gamma} = y_{1\gamma}\mu_\gamma$.

We must show that a solution $y_{1\gamma}$ to 11) exists and that $|w_\gamma'' - v_\gamma|$ is small.

If μ_1 and μ_γ have inverses and $y \in \mathbf{D}$, let

$$\sigma(y) = \left(\frac{\mu_1}{|\mu_1|}\right)^{-1} y \left(\frac{\mu_\gamma}{|\mu_\gamma|}\right) .$$

In the cases $\mathbf{D} = \mathbf{R},\mathbf{C}$, the transformation o is the identity. For $\mathbf{D} = \mathbf{Q}$, considered as a 4-dimensional vector space over $\mathbf{R}$, o is an orthogonal transformation. Equation 11) is equivalent to

$$(12) \qquad y_{1\gamma} = -\left(1 - \frac{|\mu_\gamma|}{|\mu_1|}\, o\right)^{-1} \mu_1^{-1} \left(\sum_{\beta=2}^{n} \varepsilon_{1\beta} \lambda_\beta^r y_{\beta\gamma} \right) \quad (\gamma = 2,\ldots,n)$$

in case μ_1, μ_γ and $\left(1 - \frac{|\mu_\gamma|}{|\mu_1|}\, o\right)$ have inverses.

We know that $|\mu_\gamma - \lambda_\gamma^r| < \delta |\lambda_\gamma|^r$ $(\gamma = 1,\ldots,n)$. This implies:

$$(13) \qquad \left| \frac{\mu_\gamma}{|\lambda_\gamma|^r} - 1 \right| < \delta \quad (\gamma = 1,\ldots,n),$$

$$(14) \qquad 1 - \delta < \frac{|\mu_\gamma|}{|\lambda_\gamma|^r} < 1 + \delta \quad (\gamma = 1,\ldots,n),$$

$$(15) \qquad \frac{|\mu_\gamma|}{|\mu_1|} = \frac{|\lambda_\gamma|^r}{|\lambda_1|^r} \frac{\dfrac{|\mu_\gamma|}{|\lambda_\gamma|^r}}{\dfrac{|\mu_1|}{|\lambda_1|^r}} < \frac{|\lambda_\gamma|^r}{|\lambda_1|^r} \frac{1 - \delta}{1 + \delta} \quad (\gamma = 1,\ldots,n).$$

If δ is chosen < 1, equation 14) implies that all μ_γ have inverses. Furthermore, if λ is real, $|\lambda| < 1$ then $1-\lambda o$ has an inverse and

$$(16) \qquad \|(1 - \lambda o)^{-1}\| \leq \frac{1}{1 - |\lambda|\,\|o\|} .$$

Equation 15) shows that there is an integer r_0 so that

$$(17) \qquad \frac{|\mu_\gamma|}{|\mu_1|} < \tfrac{1}{2} \quad \text{for} \quad r > r_0, \quad (\gamma = 2,\ldots,n).$$

Then for $r > r_0$, $1 - \frac{|\mu_\gamma|}{|\mu_1|}\, o$ has an inverse and since o is orthogonal, $\|o\| = 1$, and equation 16) shows that

$$(18) \qquad \left\|\left(1 - \frac{|\mu_\gamma|}{|\mu_1|}\, o\right)^{-1}\right\| \leq 2 .$$

Therefore, for $r > r_0$, equation 12) implies

$$(19) \qquad |y_{1\gamma}| \leq \left\|\left(1 - \frac{|\mu_\gamma|}{|\mu_1|}\, o\right)^{-1}\right\| \, |\mu_1^{-1}| \left| \sum_{\beta=2}^{n} \varepsilon_{1\beta} \lambda_\beta^r y_{\beta\gamma} \right| .$$

$$\leq 2 \sum |\varepsilon_{1\beta}| \frac{|\lambda_\beta|^r}{|\mu_1|} |y_{\beta\gamma}| .$$

Since $(\varepsilon_{\alpha\beta}) = x^{-1}\omega \quad x$ is contained in the compact neighborhood Ω_2, there is M so that

(20)
$$|\varepsilon_{\alpha\beta}| < M \quad .$$

Furthermore,

(21)
$$\frac{|\lambda_\beta|^r}{|\mu_1|} = \frac{|\lambda_\beta|^r}{|\lambda_1|^r}\frac{|\lambda_1|^r}{|\mu_1|} \leq \frac{|\lambda_\beta|^r}{|\lambda_1|^r}\ \frac{1}{1-\delta} \quad ,$$

and $\dfrac{|\lambda_\beta|^r}{|\lambda_1|^r} < \delta$ for large r, say $r > r_1 \geq r_0$. Since

$$|w'_\gamma - v_\gamma| = |\sum_{\beta=2}^{n} v_\beta y_{\beta\gamma} - v_\gamma \ | < \delta,$$ there is N so that

(22)
$$|y_{\beta\gamma}| \leq N \qquad (\beta,\gamma = 2,\ldots,n).$$

Then equation 19) implies for $r > r_1$

(23)
$$|y_{1\gamma}| \leq 2MN\frac{\delta}{1-\delta} \quad ,$$

(24)
$$\begin{aligned} |w''_\gamma - v_\gamma| &= |v_1 y_{1\gamma} + w'_\gamma - v_\gamma| \leq |v_1||y_{1\gamma}| \\ &+ \ |w'_\gamma - v_\gamma| < 2|v_1|\ MN\frac{\delta}{1-\delta} + \delta \\ &= K\frac{\delta}{1-\delta} + \delta \quad . \end{aligned}$$

Finally, since $x^{-1}Bx$ has the eigenvector w''_γ with the corresponding eigenvalue μ_γ $(\gamma > 1)$, it follows that B has the eigenvector $w_\gamma = x(w''_\gamma)$ with the same eigenvalue, and

$$|w_\gamma - v_\gamma| = |x(w''_\gamma) - v_\gamma| = |x(w''_\gamma) - x(v_\gamma)| \quad \|x\|\ |w''_\gamma - v_\gamma|$$

or

(25)
$$|w_\gamma - v_\gamma| < \|x\|\ (K\frac{\delta}{1-\delta} + \delta)$$

for $r > r_1$ and $\gamma = 2,\ldots,n$. Since x is compact, we conclude that if we choose δ small enough then

(26)
$$|\mu_\alpha - \lambda_\alpha^r| < \varepsilon|\lambda_\alpha|^r \qquad \alpha = 1,\ldots,n$$

(27)
$$|w_\alpha - v_\alpha| < \varepsilon \qquad \alpha = 1,\ldots,n \quad ,$$

for $r > r_1$. Lemma 1 tells us that Ω can be chosen small enough so that 26) and 27) are also valid for $r = 1,2,\ldots,r_1$.

LEMMA 4. The hypothesis here is the same as in Lemma 3. For $\varepsilon > 0$ there is a neighborhood Ω of 1 in $GL(n,\mathbf{D})$ such that (for all integers $r > 0$) if $B \in \Omega A^r \Omega^{-1}$, then B satisfies conditions 1) and 2) in Lemma 3.

PROOF. Let Ω_0 be the neighborhood of Lemma 3, with condition 2) replaced by

3) $|w_\alpha - v_\alpha| < \delta \quad (\alpha = 1,\dots,n)$.

Let Ω be a neighborhood of 1 such that

4) $\Omega^{-1}\Omega \subset \Omega_0$,

5) if $\omega \in \Omega$, then $\|\omega - 1\| < \delta$.

Let $B \in \Omega A^r \Omega^{-1}$. Then $B = \omega_1 A^r \omega_2^{-1} = \omega_2 [(\omega_2^{-1}\omega_1) A^r] \omega_2^{-1}$, where $\omega_1, \omega_2 \in \Omega$. Because of condition 4), $\omega_2^{-1}\omega_1 \in \Omega_0$, therefore $\omega_2^{-1}\omega_1 A^r$ has a representative set of eigenvalues $(\mu_1,\dots,\mu_n)$ and a corresponding eigenbasis $(w_1,\dots,w_n)$ satisfying condition 1) of Lemma 3 and the present condition 3). B has the same representative set of eigenvalues with a corresponding eigenbasis $(\omega_2(w_1), \omega_2(w_2),\dots, \omega_2(w_n))$.

Furthermore, since $\|\omega_2 - 1\| < \delta$ and $|w_\alpha| < |v_\alpha| + \delta$

$$6) \quad |\omega_2(w_\alpha) - v_\alpha| \le |\omega_2(w_\alpha) - w_\alpha| + |w_\alpha - v_\alpha|$$

$$\le \|\omega_2 - 1\| \, |w_\alpha| + \delta$$

$$< \delta(|v_\alpha| + \delta) + \delta \quad .$$

Thus if δ is small enough, then Ω satisfies the required conditions.

We recall some facts about Haar measure (which are proved in A. Weil [1]). A locally compact group is called unimodular if every left invariant Haar measure is right invariant. If G is a locally compact group and Γ is a closed, unimodular subgroup then there is a relatively invariant measure μ on the factor space $\Gamma\backslash G$. That is, there is a homomorphism $\chi : G \longrightarrow$ positive real numbers, such that for any measurable set $\Omega \subset \Gamma\backslash G$, $\mu(\Omega g) = \mu(\Omega)\, \chi(g)$. This is true in particular, if Γ is discrete. We shall call a relatively invariant measure a Haar measure on $\Gamma\backslash G$. If $\Gamma\backslash G$ has finite Haar measure, then the measure is invariant. For putting $\Omega = \Gamma\backslash G$, we have $\Omega = \Omega g$ and $\mu(\Omega) = \mu(\Omega g) = \mu(\Omega)\, \chi(g)$. Therefore $\chi(g) = 1$.

The following lemma is due to A. Selberg [1].

LEMMA 5 (Selberg). Let G be a locally compact group and let Γ be a discrete subgroup such that $\Gamma\backslash G$ has finite Haar measure. Then for any element $g \in G$ and any neighborhood Ω of 1 there is a positive integer n and $\omega_1, \omega_2 \in \Omega$ such that $\omega_1 g^n \omega_2^{-1} \in \Gamma$.

PROOF: Let $\pi : G \longrightarrow \Gamma\backslash G$ be the projection, which maps each element into its Γ-coset. Since $\Gamma\backslash G$ has finite measure and the sets $\pi(\Omega), \pi(\Omega)g,\dots, \pi(\Omega)g^n$ have the same positive measure, it follows that there are distinct integers r, s (say $r < s$) so that $\pi(\Omega)g^r \cap \pi(\Omega)\, g^s \neq 0$.

This means that there are $\omega_1, \omega_2 \in \Omega$ and $\gamma_1, \gamma_2 \in \Gamma$ so that $\gamma_1 \omega_1 g^r = \gamma_2 \omega_2 g^s$. Then $\gamma_2^{-1} \gamma_1 = \omega_2 g^{s-r} \omega_1^{-1}$.

The above lemma is the only place where we shall use the discreteness of Γ. Note that the lemma is true under the weaker assumption that Γ is closed and unimodular. However A. Borel [1] has shown that if G is non-compact, simple, Γ is closed, unimodular and $\Gamma\backslash G$ has finite Haar measure, then Γ is either the entire group G or discrete. Our feeling is that this is probably true for strongly transitive groups. For this reason we shall suppose that Γ is discrete.

In the following lemmas G is strongly transitive and Γ is a discrete subgroup with compact factor space $\Gamma\backslash G$. In particular $\Gamma\backslash G$ has finite measure, so that Lemma 5 is valid.

LEMMA 6. If ρ is any ray, $\Gamma\rho$ is dense in the space of rays.

PROOF. Let Ω be an open set of rays. We must show that there is $\gamma \in \Gamma$ so that $\gamma\rho \in \Omega$.

By ST 3 (condition 3 in the definition of strongly transitive groups) there is a semi-simple element $g_0 \in G$ which has a representative set of eigenvalues $(\lambda_1, \ldots, \lambda_n)$ with distinct absolute values. We shall suppose that $|\lambda_1| = \max_{1 \leq \alpha \leq n} |\lambda_\alpha|$. Let $(v_1, \ldots, v_n)$ be a corresponding eigenbasis and let $x = v_1 + v_2 + \ldots + v_n$. Let $\rho(x)$ be the ray through x and let $\Omega'(\varepsilon)$ be the following ray-neighborhood of $\rho(x)$:

$$\Omega'(\varepsilon) = \left\{ \rho(v) \mid v = \sum_{\alpha=1}^{n} v_\alpha d_\alpha, \ |d_\alpha - 1| < \varepsilon, \ \alpha = 1, \ldots, n \right\} .$$

By ST 2a there is $g \in G$ so that $g\rho(v_1) \in \Omega$ and $g^{-1}\rho \in \Omega'(\varepsilon)$. The element $g_1 = gg_0g^{-1}$ has a representative set of eigenvalues $(\lambda_1, \ldots, \lambda_n)$ with a corresponding eigenbasis $v_1' = g(v_1), \ldots, v_n' = g(v_n)$. $g^{-1}\rho = \rho(v)$ where $v = \sum_{\alpha=1}^{n} v_\alpha d_\alpha$ and $|d_\alpha - 1| < \varepsilon$, $\alpha = 1, \ldots, n$. This means that $\rho = \rho(v')$, where $v' = \sum_{\alpha=1}^{n} v_\alpha' d_\alpha$.

Suppose that we change the basis from $v_1', \ldots, v_n'$ to $w_1', \ldots, w_n'$. Then $v' = \sum_{\alpha=1}^{n} v_\alpha' d_\alpha = \sum_{\alpha=1}^{n} w_\alpha' f_\alpha$. For $\varepsilon > 0$ there is $\delta > 0$ so that

1) $|f_\alpha - d_\alpha| < \varepsilon, \quad \alpha = 1, \ldots, n,$

when

$$|w_\alpha' - v_\alpha'| < \delta, \quad \alpha = 1, \ldots, n.$$

If δ is small enough, $|w_1' - v_1'| < \delta$ guarantees that $\rho(w_1') \in \Omega$. Lemmas 4 and 5 imply that there is a semi-simple element $\gamma \in \Gamma$ which has a representative set of eigenvalues $(\mu_1,\ldots,\mu_n)$ and a corresponding eigenbasis $(w_1',\ldots,w_n')$ such that

2) $|w_\alpha' - v_\alpha'| < \delta \qquad (\alpha = 1,\ldots,n)$,

3) $|\mu_1| > |\mu_\alpha| \qquad (\alpha = 2,\ldots,n)$.

Let us now compute $\gamma^m\rho$. $\rho = \rho(v')$, where $v' = \sum_{\alpha=1}^{n} w_\alpha' f_\alpha$

and $|f_\alpha - 1| \leq |f_\alpha - d_\alpha| + |d_\alpha - 1| < 2\varepsilon$.

4) $\gamma^m v' = \sum_{\alpha=1}^{n} w_\alpha' \mu_\alpha^m f_\alpha$.

Now $\gamma^m\rho = \gamma^m\rho(v') = \gamma^m\rho\left(v' \frac{1}{|\mu_1|^m}\right) = \rho\left[\gamma^m(v') \frac{1}{|\mu_1|^m}\right]$

and

5) $\gamma^m(v') \frac{1}{|\mu_1|^m} = \sum_{\alpha=1}^{n} w_\alpha' \frac{\mu_\alpha^m}{|\mu_1|^m} f_\alpha$.

Since $\frac{\mu_\alpha^m}{|\mu_1|^m} \longrightarrow 0$ as $m \longrightarrow \infty$ $(\alpha > 1)$, for large m, we have

6) $\left|\gamma^m(v') \frac{1}{|\mu_1|^m} - w_1' \frac{\mu_1^m}{|\mu_1|^m} f_1\right| < \varepsilon$.

In case $\mathbf{D} = \mathbf{R}$, f_1 and μ_1 are real and (for small enough ε) 6) guarantees that $\gamma^m\rho \in \Omega$. For $\mathbf{D} = \mathbf{C}$ or $\mathbf{Q}$, μ_1 can be represented as

7) $\mu_1 = a + bh$,

where $a,b \in R$ and h is pure imaginary with $|h| = 1$. There is an angle θ so that $a = |\mu_1| \cos\theta$, $b = |\mu_1| \sin\theta$ and

8) $\mu_1 = |\mu_1|(\cos\theta + h \sin\theta)$

9) $\mu_1^m = |\mu_1^m|(\cos m\theta + h \sin m\theta)$.

If θ is a rational multiple of π, then $\mu_1^m = |\mu_1|^m$ for some m. For large enough r, $\gamma^{mr}\rho \in \Omega$. If θ is an irrational multiple of π, then there is a sequence $m_i \longrightarrow \infty$ such that $\cos m_i\theta + h \sin m_i\theta \longrightarrow 1$. For large enough m_i (and small ε), $\gamma^{m_i}\rho \in \Omega$.

The following lemma is due to Hedlund [1].

LEMMA 7 (Hedlund). Let Ω be an open set in V. Suppose that for any (non-empty) open subset $\Sigma \subset \Omega$, $\Gamma\Sigma$ is dense in V. Then Ω contains a point v such that Γv is dense in V.

PROOF. Let $\Omega_1, \Omega_2, \ldots$ be a countable base for the open sets in V. Since $\Gamma\Omega$ is dense in V, there is a point $v_1 \in \Omega$ which has a Γ-image in Ω_1. By continuity, there is a closed ball $B_1 = \{v \mid |v - v_1| \leq r_1\} \subset \Omega$, such that all points in B_1 have Γ-images in Ω_1. Since B_1 contains an open subset of Ω, ΓB_1 is dense in V. Therefore there is a point $v_2 \in B_1$ which has a Γ-image in Ω_2. Then there is a closed ball $B_2 \subset B_1$ such that all points in B_2 have Γ-images in Ω_1 and Ω_2. Continuing in this way, we can find a sequence of closed balls $B_1 \supset B_2 \supset \dots \supset B_n \supset \dots$, such that all points in B_n have Γ-images in $\Omega_1, \Omega_2, \dots, \Omega_n$. $\bigcap_{n=1}^{\infty} B_n$ contains at least one point $v \in \Omega$, and it is clear that Γv is dense in V.

LEMMA 8. There exists a vector $v \in V$ such that Γv is dense in V.

PROOF. A subset S in V will be called convex if $s_1, s_2 \in S$ implies that the segment $[s_1, s_2] = \{s \mid s = s_1(1-\lambda) + s_2\lambda, \lambda \text{ real}, 0 \leq \lambda \leq 1\} \subset S$. We shall show that if Ψ is a convex open set in $V - \{0\}$, then $\Gamma\Psi$ is dense in V. The lemma will then follow from Lemma 7.

Let Ω be an open set in V. We must show that there exists $\gamma \in \Gamma$ and $v \in \Psi$ such that $\gamma(v) \in \Omega$.

Let Ψ_ρ and Ω_ρ be the following open sets of rays:

$$\Psi_\rho = \{\rho(v) \mid v \in \Psi\} ,$$

$$\Omega_\rho = \{\rho(v) \mid v \in \Omega\} ,$$

and let $\rho_0 \in \Omega_\rho$. By Lemma 6 there is $\gamma_0 \in \Gamma$ so that $\gamma_0\rho_0 \in \Psi_\rho$. Since γ_0 is continuous, there is an open set of rays $\Sigma_\rho \subset \Omega_\rho$ such that γ $\gamma_0 \Sigma_\rho \subset \Psi_\rho$.

By ST 3 there is an element $g \in G$ which has a representative set of eigenvalues $(\lambda_1, \dots, \lambda_n)$ such that

1) $|\lambda_1|, \dots, |\lambda_n|$ are distinct,

2) $|\lambda_1| > 1 > |\lambda_2|$.

Let $(v_1, \dots, v_n)$ be a corresponding eigenbasis for g. By ST b) there is an element $h \in G$ so that $h\rho(v_1)$, $h\rho(v_2) \in \Sigma_\rho$. The element $k = hgh^{-1} \in G$ has the same representative set of eigenvalues $(\lambda_1, \dots, \lambda_n)$ with a corresponding eigenbasis $(h(v_1), h(v_2), \dots, h(v_n))$. By Lemmas 4 and 5, there is a semi-simple element $\gamma_1 \in \Gamma$ which has a representative set of eigenvalues $(\mu_1, \dots, \mu_n)$ and a corresponding eigenbasis $(w_1, \dots, w_n)$ so that

3) $|\mu_\alpha - \lambda_\alpha^r| < \varepsilon |\lambda_\alpha|^r \qquad (\alpha = 1, \dots, n)$

4) $|w_\alpha - h(v_\alpha)| < \varepsilon \qquad (\alpha = 1, \dots, n)$,

where ε is arbitrary, and $r = r(\varepsilon)$ is a positive integer depending on ε. For small ε, condition 3) implies that $|\mu_1| > 1 > |\mu_2|$, and condition 4) implies that $\rho(w_1), \rho(w_2) \in \Sigma_\rho$.

Since $\gamma_0 \Sigma_\rho \subset \Psi_\rho$, $\gamma_0\rho(w_1)$ and $\gamma_0\rho(w_2) \in \Psi_\rho$. This means that there is a point $p_1 \in \gamma_0\rho(w_1) \cap \Psi$ and a point $p_2 \in \gamma_0\rho(w_2) \cap \Psi$. The segment $[p_1,p_2] = \{p \mid p = p_1(1-\lambda) + p_2\lambda,\ \lambda \text{ real},\ 0 \leq \lambda \leq 1\}$ is contained in Ψ, since Ψ is convex. The segment $\gamma_0^{-1}[p_1,p_2] = [q_1,q_2]$, where $q_1 = \gamma_0^{-1}(p_1) \in \rho(w_1)$ and $q_2 = \gamma_0^{-1}(p_2) \in \rho(w_2)$.

Since $q_\alpha \in \rho(w_\alpha)$ $(\alpha = 1,2)$, $q_\alpha = w_\alpha r_\alpha$, where r_α is real and > 0. Therefore $\gamma_1 q_\alpha = q_\alpha r_\alpha^{-1}\lambda_\alpha r_\alpha = q_\alpha\lambda_\alpha$. Thus $\gamma_1^r [q_1,q_2] = [q_1\lambda_1^r, q_2\lambda_2^r]$. Since $|\lambda_2| < 1$, $q_2\lambda_2^r \longrightarrow 0$ as $r \longrightarrow \infty$. Since $|\lambda_1| > 1$, $|q_1\lambda_1^r| \longrightarrow \infty$ as $r \longrightarrow \infty$. λ_1 can be represented in the form $\lambda_1 = a + ub$, where a and b are real, and u is pure imaginary of norm 1 (if $D = R$, then $b = 0$). Since $|\lambda_1|^2 = a^2 + b^2$, there is θ such that $a = |\lambda_1| \cos\theta$ and $b = |\lambda_1| \sin\theta$. Then $\lambda_1 = |\lambda_1|(\cos\theta + u \sin\theta)$, and $\lambda_1^r = |\lambda_1|^r (\cos r\theta + u \sin r\theta)$. There is always a sequence $r_k \longrightarrow \infty$ such that $\cos r_k\theta + u \sin r_k\theta \longrightarrow 1$. The segment $\gamma_1^{r_k}[q_1,q_2] = [q_1\lambda_1^{r_k}, q_2\lambda_2^{r_k}]$ approaches the ray $\rho(w_1)$. Therefore, for large r_k, $\gamma_1^{r_k}[q_1,q_2]$ intersects Ω. Thus there is a point $p \in [p_1,p_2] \subset \Psi$ such that $\gamma_1^{r_k}\gamma_0^{-1}p \in \Omega$.

LEMMA 9. Let v be an eigenvector of an element $\gamma \in \Gamma$, such that the corresponding eigenvalue λ has norm $|\lambda| \neq 1$. Then Γv is dense in V.

PROOF. By the previous lemma, there is $w \in V$ so that Γw is dense in V. It is easily seen that if $d \in \mathbf{D}$, $d \neq 0$, then $\Gamma(wd)$ is dense in V.

By Lemma 6 there is a sequence $\{\gamma_i\} \subset \Gamma$ such that $\gamma_i\rho(v) \longrightarrow \rho(w)$. Now $\gamma_i\gamma^r(v) = (\gamma_i v)\lambda^r$. For each i there is an integer r_i so that $1 \leq |\gamma_i v||\lambda|^{r_i} \leq |\lambda|$. The sequence of vectors $\{\gamma_i\gamma^{r_i}v\}$ has a subsequence $\{\gamma_j\gamma^{r_j}v\}$ which converges to a vector $u \neq 0$.

Let w_j be the vector on the ray $\gamma_j\rho(v)$ such that $|w_j| = |w|$. Since $\gamma_j\rho(v) \longrightarrow \rho(w)$, it follows that $w_j \longrightarrow w$. If $[x]$ denotes the subspace spanned by x, then $[\gamma_j\gamma^{r_j}v] = [\gamma_j v] = [w_j]$. Thus for each j there is $d_j \in \mathbf{D}$ so that $\gamma_j\gamma^{r_j}v = w_j d_j$. Since $1 \leq |w_j d_j| \leq |\lambda|$ and and $|w_j d_j| = |w||d_j|$, the sequence $\{d_j\}$ has a subsequence $\{d_k\}$ which converges to an element $d \in \mathbf{D}$, $d \neq 0$. We now have $w_k d_k \longrightarrow u$, $w_k \longrightarrow w$ and $d_k \longrightarrow d$. It follows that $u = wd$. Since Γw is dense in V, the same is true of Γu. Since a sequence of Γ-images of v converges to u, and Γu is dense in V, it now follows that Γv is dense in V.

LEMMA 10. If $v \neq 0$ and there is a sequence $\{\gamma_i\} \subset \Gamma$ so that $\lim_{i \to \infty} \gamma_i v = 0$, then Γv is dense in V.

PROOF. The sequence of rays $\{\gamma_i \rho(v)\}$ has a subsequence $\{\gamma_j \rho(v)\}$ which converges to a ray ρ_0.

By ST 3 there is a semi-simple element $g \in G$ which has a representative set of eigenvalues $(\lambda_1, \ldots, \lambda_n)$ such that

1) $|\lambda_1|, \ldots, |\lambda_n|$ are distinct

2) $|\lambda_1| > 1$.

Let $(u_1, \ldots, u_n)$ be a corresponding eigenbasis, let $u = u_1 + \ldots + u_n$ and let $\Omega_{u,\delta} = \{\rho(x) \mid x = \sum_{\alpha=1}^{n} u_\alpha d_\alpha, \ |d_\alpha - 1| < \delta\}$. By ST 2a there is $g_1 \in G$ so that $g_1^{-1} \rho_0 \in \Omega_{u,\delta}$. The element $g_2 = g_1 g g_1^{-1}$ has a representative set of eigenvalues $(\lambda_1, \ldots, \lambda_n)$ and a corresponding eigenbasis $(w_1, \ldots, w_n)$ where $w_\alpha = g_1(u_\alpha)$, $\alpha = 1, \ldots, n$. The ray $\rho_0 = \rho(w)$, where $w = \sum_{\alpha=1}^{n} w_\alpha d_\alpha$ and $|d_\alpha - 1| < \delta$. By Lemmas 4 and 5 there is a semi-simple element $\gamma \in \Gamma$ which has a representative set of eigenvalues $(\mu_1, \ldots, \mu_n)$ and a corresponding eigenbasis $(x_1, \ldots, x_n)$ so that

3) $|\mu_1|, \ldots, |\mu_n|$ are distinct,

4) $|\mu_1| > 1$

5) $|x_\alpha - w_\alpha| < \delta \qquad (\alpha = 1, \ldots, r)$.

$$w = \sum_{\alpha=1}^{n} w_\alpha d_\alpha = \sum_{\alpha=1}^{n} x_\alpha f_\alpha, \text{ where}$$

6) $|f_\alpha - d_\alpha| < \varepsilon$

when δ is small. This implies in particular (for small ε) that $f_\alpha \neq 0$. We may suppose that the eigenvalues $(\mu_1, \ldots, \mu_n)$ are ordered so that $|\mu_1| > |\mu_\alpha| \quad (\alpha = 2, \ldots, n)$.

Let $\gamma_j v = \sum_{\alpha=1}^{n} x_\alpha f_{\alpha j}$, and let $v_j \in \gamma_j \rho(v)$ with $|v_j| = |w|$. Since $\gamma_j \rho(v) \longrightarrow \rho(w)$, it follows that $v_j \longrightarrow w$. Since v_j and $\gamma_j v$ are on the same ray, $\gamma_j v = v_j r_j$, where r_j is real and > 0. If $v_j = \sum_{\alpha=1}^{n} x_\alpha f'_{\alpha j}$, then $f_{\alpha j} = f'_{\alpha j} r_j$ and $f'_{\alpha j} \longrightarrow f_\alpha$ as $j \longrightarrow \infty$ $(\alpha = 1, \ldots, n)$. This means that $\frac{|f_{\alpha j}|}{|f_{\beta j}|} \longrightarrow \frac{|f_\alpha|}{|f_\beta|}$ as $j \longrightarrow \infty$. Also, since $\gamma_j v \longrightarrow 0$, $|f_{\alpha j}| \longrightarrow 0$ as $j \longrightarrow \infty$ $(\alpha = 1,\ldots,n)$.

Since $\frac{f_{1j}}{r_j} \longrightarrow f_1 \neq 0$, $f_{1j} \neq 0$ for large j. Since also $f_{1j} \longrightarrow 0$, we have $0 < |f_{1j}| < 1$ for large j. We may suppose that this is true for all j. Then for each j there is a positive integer s_j so that $1 \leq |f_{1j}|\ |\mu_1|^{s_j} \leq |\mu_1|$. The sequence $\{f_{1j}\mu_1^{s_j}\}$ has a subsequence $\{f_{1k}\mu_1^{s_k}\}$ which converges to some element $f \in \mathbf{D}$, $f \neq 0$.

We now consider the sequence $\{\gamma^{s_k}\ \gamma_k v\}$.

$$\gamma^{s_k}\ \gamma_k v = \sum_{\alpha=1}^{n} x_\alpha f_{\alpha k}\mu_\alpha^{s_k} .$$

For $\alpha \neq 1$,

$$|f_{\alpha k}\mu_\alpha^{s_k}| = \frac{|f_{\alpha k}|}{|f_{1k}|}\ |f_{1k}\mu_1^{s_k}|\ \frac{|\mu_\alpha|^{s_k}}{|\mu_1|^{s_k}} .$$

Since $\frac{|f_{\alpha k}|}{|f_{1k}|} \longrightarrow \frac{|f_\alpha|}{|f_1|}$, $|f_{1k}\mu_1^{sk}| \longrightarrow |f|$ and $\frac{|\mu_\alpha|^{s_k}}{|\mu_1|^{s_k}} \longrightarrow 0$ as $k \longrightarrow \infty$, it follows that $f_{\alpha k}\mu_\alpha^{s_k} \longrightarrow 0$ as $k \longrightarrow \infty$. Thus $\gamma^{s_k}\ \gamma_k v \longrightarrow x_1 f$ as $k \longrightarrow \infty$.

By Lemma 9, $\Gamma(x_1 f)$ is dense in V. Since a sequence of Γ-images of v converges to $x_1 f$, it now follows that Γ_v is dense in V.

LEMMA 11. The transitivity of G and the compactness of $\Gamma\backslash G$ imply that for any $v \in V$ there is a sequence $\{\gamma_i\} \subset \Gamma$ so that $\lim_{i \to \infty} \gamma_i v = 0$.

PROOF. Let $v \neq 0$. Since G is transitive, $Gv = V - \{0\}$. There is a compact subset $K \subset G$ so that $G = \Gamma K$. The set Kv is a compact subset of V which has the property that for any $w \in V - \{0\}$, there is $\gamma \in \Gamma$ with $\gamma w \in Kv$. Since Kv is compact, there is a number M so that if $w \in Kv$, then $|w| < M$.

Now let $\{\lambda_k\}$ be a sequence in $\mathbf{D}$ so that $|\lambda_j| \longrightarrow \infty$ as $j \longrightarrow \infty$. For each j there is $\gamma_j \in \Gamma$ so that $\gamma_j(v\lambda_j) \in Kv$. Then $|\gamma_j(v)|\,|\lambda_j| = |\gamma_j(v\lambda_j)| < M$, and $|\gamma_j(v)| < \frac{M}{|\lambda_j|}$. Therefore $|\gamma_j(v)| \longrightarrow 0$ and $\gamma_j(v) \longrightarrow 0$ as $j \longrightarrow \infty$.

§3. Geodesic Flows. The usual geodesic flow on a Riemannian manifold M is a 1-parameter group φ_t of transformations of the space U of unit tangent vectors on M. If $u \in U$, $\varphi_t u$ is obtained by parallel translating the vector u distance t along the oriented geodesic which is tangent to u. The properties of this flow on locally symmetric spaces have been studied (cf. G. Hedlund [2], E. Hopf [2], F. Mautner [1]).

We shall consider here an extension of the flow to the space F of orthonormal n-frames on the manifold M. A frame $(v_1, \ldots, v_n)$ is parallel translated distance t along the geodesic determined by v_1.

We shall consider the particular case of a compact 3-dimensional manifold M with constant negative curvature. Such a manifold has as universal covering space the 3-dimensional hyperbolic space H_3. The isometry group of H_3 is $G = SL(2,\mathbf{C})$ modulo its center. If $\Gamma \subset G$ is the group of covering transformations corresponding to M, the space of frames F may be identified with $\Gamma\backslash G$, and the geodesic flow in F is given by $\Gamma g \longrightarrow \Gamma g h_t$, where

$$h_t = \begin{pmatrix} e^t & 0 \\ 0 & e^{-t} \end{pmatrix}$$

DEFINITION: A flow φ_t on a space F is called permanently regionally transitive if for any two open subsets Ω, Σ in F, there is a number t_0, so that

$$\varphi_t(\Omega) \cap \Sigma \neq \emptyset \quad \text{for} \quad t > t_0.$$

THEOREM 3. Let M be a 3-dimensional compact manifold with constant negative curvature, and let F be the space of 3-frames on M. The geodesic flow in F is permanently regionally transitive.

PROOF. Let Γ be the group of covering transformations corresponding to M. Γ is a discrete subgroup of $G = SL(2,\mathbf{C})$ with compact factor space $\Gamma\backslash G$. We must show that if Ω and Σ are any two open subsets of G, there is a number t_0 so that

$$1) \quad \Gamma\Omega h_t \cap \Sigma \neq \emptyset \quad \text{for} \quad t > t_0, \quad \text{where} \quad h_t = \begin{pmatrix} e^t & 0 \\ 0 & e^{-t} \end{pmatrix}.$$

Note that it is sufficient to show this for the case that Ω is a neighborhood of the identity. Any open set contains a set $g\Omega$ where Ω is a neighborhood of the identity and $\Gamma g\Omega h_t = g(g^{-1}\Gamma g)\Omega h_t$. If Γ has compact factor space, so has $g^{-1}\Gamma g$. If we have shown 1) for any discrete Γ with compact factor space and any neighborhood Ω of 1, then it follows that $(g^{-1}\Gamma g)\Omega h_t$ intersects $g^{-1}\Sigma$ for $t > t_0$, therefore

$$2) \quad \Gamma g\Omega h_t \cap \Sigma \neq \emptyset \quad \text{for} \quad t > t_0.$$

Thus it is enough to prove 1) for neighborhoods Ω of the identity.

Let P be the subgroup of G consisting of all matrices of the form

$$\begin{pmatrix} 1 & 0 \\ a & 1 \end{pmatrix}$$

P is the subgroup of G which leaves the vector $(0,1)$ fixed. Every isotropy group (of a vector in the 2-dimensional complex vector space V) is of the form gPg^{-1} for some $g \in G$. The theorem of the previous section states that Γv is dense in V, for $v \neq 0$. This implies that $\Gamma g P g^{-1}$ is dense in G, for any $g \in G$.

$$\text{If } p = \begin{pmatrix} 1 & 0 \\ a & 1 \end{pmatrix} \in P, \text{ then } h_t\, p\, h_t^{-1} = \begin{pmatrix} 1 & 0 \\ e^{-2t}a & 1 \end{pmatrix}.$$

This shows that for any $p \in P$ and any neighborhood Ω of 1, there is a number $t_0 = t_0(p,\Omega)$ so that

$$3) \quad h_t\, p\, h_t^{-1} \in \Omega \quad \text{for } t > t_0.$$

Since $\Gamma \backslash G$ is compact, there is a compact set $K \subset G$ so that for any $g \in G$ there is $\gamma \in \Gamma$ with $\gamma g \in K$. For any $k \in K$, $\Gamma k P k^{-1}$ is dense in G, therefore there is $\gamma_k \in \Gamma$ and $p_k \in P$ so that $\gamma_k k p_k k^{-1} \in \Sigma k^{-1}$, or $\gamma_k k p_k \in \Sigma$. By continuity, there is a neighborhood $\Omega(k)$ of k such that

$$4) \quad \gamma_k \Omega(k) p_k \cap \Sigma \neq \emptyset.$$

K is covered by a finite number of these neighborhoods: $\Omega(k_1),\ldots,\Omega(k_n)$. We shall denote γ_{k_i} by γ_i and p_{k_i} by p_i $(i = 1,\ldots,n)$. Then if $k \in K$,

$$5) \quad \gamma_i\, k p_i \in \Sigma \text{ for some } i.$$

By 3) there is t_0 so that

$$6) \quad h_t\, p_i\, h_t^{-1} = \omega_{i,t} \in \Omega, \quad \text{for } t > t_0,\ i = 1,\ldots,n.$$

For each t, there is $\gamma \in \Gamma$ so that $\gamma_t\, h_t = k_t \in K$. By 5), for each t there is some i so that $\gamma_i \gamma_t\, h_t\, p_i = \gamma_i\, k_t\, p_i \in \Sigma$. Then for $t > t_0$,

$$\gamma_i\, \gamma_t\, \omega_{i,t}\, h_t = \gamma_i\, \gamma_t\, h_t\, h_t^{-1}\, \omega_{i,t}\, h_t = \gamma_i\, \gamma_t\, h_t\, p_i \in \Sigma.$$

Thus $\Gamma \Omega\, h_t \cap \Sigma \neq \emptyset$ for $t > t_0$.

BIBLIOGRAPHY

Auslander, L.,

1) Fundamental Groups of Compact Solvmanifolds, Amer. Jour. Math., Vol. 82 (4) 1960, pp. 689-697.

2) Solvable Groups Acting on Nilmanifolds, Amer. Jour. Math., Vol. 82 (4) 1960, pp. 653-660.

Auslander, L., Hahn, F., Markus, L., Topological Dynamics on Nilmanifolds, Bull. Amer. Math. Soc., Vol. 67 (1961), pp. 298-299.

Borel, A., Density Properties for Certain Subgroups of Semi-Simple Groups without Compact Components, Ann. of Math. 72 (1960) pp. 179-188.

Borel, A., and Hirzebruch, F., Characteristic Classes and Homogeneous Spaces, I. Amer. Jour. Math. 80 (1958), pp. 458-538.

Bourbaki, N., Groupes et Algèbres de Lie, Paris 1960.

Bruhat, F., Sur les Représentations Induites des Groupes de Lie, Bull. Soc. Math. de France 84 (1956) pp. 97-205.

Cartan, H., and Eilenberg, S., Homological Algebra, Princeton University Press, 1956.

Chevalley, C.,

1) Theory of Lie Groups, Princeton, 1946.

2) On the Topological Structure of Solvable Groups, Ann. of Math. Vol. 42 (1941).

Cohn, P. M., Lie Groups, Cambridge University Press, 1957.

van der Corput, J. G., Rhythmic System. Indagationes Math. Ned Akad. van Wet. Proc. of the Sect. of Sciences, Vol. 8, Fas. 4, pp. 416-429.

Dieudonné, Les Déterminants sur un corps non Commutatif, Bull. Soc. Math. France, 71, (1943), pp. 27-45.

Dixmier, J.,

1) Sur les Représentations Unitaires des Groupes de Lie Nilpotents, V, Bull. Soc. Math. France, Vol. 87 (1959) pp. 65-79.

2) L'application Exponentielle dans les Groupes de Lie Résolubles, Bull. Soc. Math. France, Vol. 85 (1957) pp. 113-121.

Ellis, R., Distal Transformation Groups, Pac. J. Math. 8, pp. 401-405, (1958).

Fomin, S., On Dynamical Systems with Pure Point Spectrum, Doklady Akad. Nauk SSSR, Vol. 77 (1951) pp. 29-32.

Gelfand, I. M., and Fomin, S. V., Geodesic Flows on Manifolds of Constant Negative Curvature, Uspehi Mat. Nauk 7 (1952) no. 1 (47), pp. 118-137 (Amer. Math Soc. Translations Series 2, Vol. 1 (1955), 49-66).

Gottschalk, W., and Hedlund, G., Topological Dynamics, A.M.S. Col. Vol. 38, New York (1955).

Green, L., Spectra of Nilflows, Bull. Amer. Math. Soc., Vol. 67 (1961), pp. 414-415.

Hedlund, G. A.,

1) Fuchsian Groups and Transitive Horocycles, Duke Jour. of Math., Vol. 2 (1936) pp. 530-542.

2) The Dynamics of Geodesic Flows, Bull. Amer. Math. Soc., 45 (1939), pp. 241-260.

Hopf, E.,

1) Ergodentheorie, Ergebnisse der Math. Vol. 2, J. Springer, Berlin, 1937. (Chelsea, New York, 1948.)

2) Statistik der Geodätischen Linien in Mannigfaltigkeiten Negativer Krümmung. Ber. Verh. Sachs. Akad. Wiss. Leipzig, Math.- Nat. Kl. Vol. 51 (1939) 261-304.

Iwasawa, K., On Some Types of Topological Groups, Ann. of Math. (2) 50 (1949) pp. 507-557.

Jacobson, N., Lie Algebras, Interscience, New York, 1962.

Kneser, H., Reguläre Kurvenscharen auf den Ringflächen, Math. Annalen, Vol. 91 (1924) pp. 135-154.

Koksma, J., Diophantische Approximationen, Berlin 1936.

Mackey, G. W.,

1) The Theory of Group Representations, Chicago 1955, (mimeographed).

2) Induced Representations of Locally Compact Groups I, Ann. of Math. 55 (1952) 101-139.

3) Unitary Representations of Group Extensions I, Acta. Math. 99 (1958) pp. 265-311.

Mahler, K., An Arithmetic Property of Groups of Linear Transformations, Acta. Arith. 5 (1959) pp. 197-203.

Malcev, A., On a Class of Homogeneous Spaces, Izvestiya Akad. Nauk SSSR Ser., Mat 13, pp. 9-32 (1949). Am. Math. Soc. Translation 39 (1949).

Matsushima, Y., On the Discrete Subgroups and Homogeneous Spaces on Nilpotent Lie Groups, Nagoya Math. Jour., 2 (1951) 95-110.

Mautner, F., Geodesic Flows on Symmetric Riemann Spaces, Ann. of Math. Vol. 65 (1957) pp. 416-431.

Milnor, J., The Geometric Realization of a Semi-simplicial Complex, Ann. Math. 65 (1957), 357-362.

Montgomery, D., and Zippin, L., Topological Transformation Groups, Interscience, N. Y., 1955.

Mostow, G. D.,

1) Factor Spaces of Solvable Groups, Ann. of Math., Vol. 60 (1954) 1-27.

2) Homogeneous Spaces with Finite Invariant Measures, Ann. of Math., Vol. 75 (1962) pp. 17-38.

Naimark, M. A., Normed Rings, Moscow 1956 (Groningen 1959).

Nielsen, J., Nogle grundlaeggende begreber vedrørende diskontinuerte grupper af lineaere substitutioner i en kompleks veriabel, Den 11te Skandinaviske Matematiker-kongress i Trondheim (1949), pp. 61-70.

Nemytskii, V. V., and Stepanov, V. V., Qualitative Theory of Differential Equations, Princeton, 1960.

Parasyuk, O. S., Horocycle Flows on Surfaces of Constant Negative Curvature, Uspehi Mat. Nauk 8 (1953) No. 3 (55), pp. 125-126.

Poincaré, H., Sur les Courbes Définies par les Équations Differentielles, Jour. de Math. (4), Vol. 1 (1885).

Rohlin, V., New Progress in the Theory of Transformations with Invariant Measures, Uspehi Mat. Nauk 15 (1960) No. 4 (94) pp. 3-26.

Saito, M., Sur Certains Groupes de Lie Résolubles, Scientific Papers of the Coll. of Gen. Ed., University of Tokyo 7 (1957); I 1-11, II 157-168.

Schwartzman, S., Asymptotic Cycles, Ann. of Math., Vol. 66 No. 2 (1957) pp. 270-284.

Segal, I. E., and von Neumann, J., A Theorem on Unitary Representations of Semi-Simple Lie Groups, Ann. of Math 52 (1950) pp. 509-517.

Seifert, H., Topologie dreidimensionaler gefaserter Räume, Acta Math. 60 (1932), pp. 147-238.

Seifert, H., and Threlfall, W., Lehrbuch der Topologie, Teubner, Leipzig 1934.

Selberg, A., On discontinuous groups in higher dimensional symmetric spaces, Contributions to Function Theory, Tata Inst. of Fund. Research, Bombay (1960).

Serre, J.-P., Cohomologie Modulo 2 des Complexes d'Eilenberg-MacLane, Comm. Math. Helv. 27 (1953), pp. 198-231.

Takenouchi, O., Sur la facteur-représentation d'un groupe de Lie résoluble de type (E), Math. Jour. Okayoma University, Vol 7 (1957) pp. 151-161.

Tuller, A., The measure of transitive geodesics on certain three-dimensional manifolds, Duke Math. Jour., Vol. 4 (1938) pp. 78-94.

Wang, H. C., Discrete subgroups of solvable Lie groups I, Ann. of Math. Vol. 64 (1956) pp. 1-19.

Weil, A., L'Integration dans les groupes topologiques et ses applications, Hermann, Paris (1940).

Weyl, H., Über die gleichverteilung von Zahlen mod. Eins, Math. Ann., 77, pp. 313-352 (1916).